# Catalyzed Synthesis of
# Natural Products

# Catalyzed Synthesis of Natural Products

Special Issue Editor

**David Díez**

MDPI • Basel • Beijing • Wuhan • Barcelona • Belgrade

MDPI

*Special Issue Editor*
David Díez
Universidad de Salamanca
Spain

*Editorial Office*
MDPI
St. Alban-Anlage 66
4052 Basel, Switzerland

This is a reprint of articles from the Special Issue published online in the open access journal *Catalysts* (ISSN 2073-4344) from 2018 to 2019 (available at: https://www.mdpi.com/journal/catalysts/special_ issues/natural_products).

For citation purposes, cite each article independently as indicated on the article page online and as indicated below:

LastName, A.A.; LastName, B.B.; LastName, C.C. Article Title. *Journal Name* **Year**, *Article Number*, Page Range.

**ISBN 978-3-03921-948-3 (Pbk)**
**ISBN 978-3-03921-949-0 (PDF)**

Cover image courtesy of David Díez Martín.

# Contents

# About the Special Issue Editor

**David Díez** was born in Salamanca and graduated with First Class Honors from the University of Salamanca in 1982, obtaining his doctorate in Chemistry from the University of Salamanca in 1986, with honors, under the supervision of Profs. J. G. Urones and I. S. Marcos. He continued his postdoctoral research (1988–1990, British Council Fellowship) with Prof. Steve V. Ley at Imperial College of Science, Technology and Medicine (London). He was then appointed Associate Professor in 1991 at the University of Salamanca and, later, as Full Professor in 2008. He is the co-author of more than 190 papers and has supervised more than 20 PhD theses. He has been Editor of a Special Issue of Catalysts and co-Editor of a Special Issue of Molecules. He is a referee for numerous international scientific journals (including Org. Lett., J. Org. Chem., Synlett, Tetrahedron, Catalysts, Molecules, and Arkivoc). He is a Fellow of the RCS and member of the RSEQ. His current research interests are focused on the transformation of natural products into biological active compounds, the chemistry of cyclopropanes, sulfones, tetrahydropyrans, chiral amides, and organocatalysis. He is Premio Maria de Maeztu to the Scientific Excellence and Dean of the Faculty of Chemical Sciences of the University of Salamanca since his appointment in 2016.

*catalysts*

MDPI

*Editorial*
# Catalyzed Synthesis of Natural Products

## David Diez

Department of Organic Chemistry, University of Salamanca, Castilla y León, 37008 Salamanca, Spain; ddm@usal.es

Received: 21 October 2019; Accepted: 23 October 2019; Published: 25 October 2019

Natural Products are secondary metabolites, that have been the inspiration for chemists and chemical biologists for many years and have a special relevance in the chemical space [1]. From ancient times, they have held the interest of the population due to their applications in life. Natural products can be isolated from both terrestrial or marine plants, animals, and microorganisms. Until two centuries ago, when morphine was isolated, the use of natural products was through employing extracts of plants or animals. The isolation of this natural product changed the way these compounds and their properties were observed by the scientific community. In the recent two hundred years, there has been an enormous development of their isolation, structural determination, and synthesis. The interest in the synthesis of these compounds is not only for confirming structure by spectroscopic methods but also to obtain intermediates and analogues in order to conduct structure–activity relationship (SAR) studies and increase the biological activity of the natural product itself. In the last years, several synthetic strategies have appeared such as diversity-oriented synthesis (DOS) [2], biological-oriented synthesis (BiOS) [3], and function-oriented synthesis (FOS) [4], in order to access to complex and functional diverse molecules that fill the chemical space. [5] In this manner, synthesis has evolved towards a simpler and ecological way to obtain natural products [6,7], using biotransformation [8], combinatorial chemistry [9], or catalysts [10] that improve the synthesis by diminishing the use of metals and, since 2000, also making use of organocatalysts [11].

In this issue, some of these methodologies have been recompiled in order to have an idea of the best methods for the synthesis of natural products by catalysis.

In this manner, Prof. Chojnacka, using immobilized lipases as catalyst, has obtained structured phosphatidylcholine enriched with myristic acid. [12] Profs. Vila and Pedro made use of enantioselective catalysts derived from (S)-mandelic acid for studying the catalytic enantioselective addition of dimethylzinc to isatins [13]. An extra step is the use of organocatalysts, in this manner, Prof. Diez shows the possibility of the obtention, in a diastereoselective manner, of the 7,8-carvone epoxides using organocatalysts as proline, quinidine, and diphenylprolinol [14]. Another methodology for the synthesis of natural products in a cheap, simple, clean and scalable method is the use of deep eutectic mixtures as reaction media. Profs. Alonso and Guillena tell us about the use of this methodology for the enantioselective organocatalyzed α-amination of 1,3-dicarbonyl compounds [15]. As has been said, not only synthesis but biotransformation has been one of the methodologies for more efficient synthesis of natural products. In this respect, Prof. Wu illustrates us with the biotransformation of ergostane triterpenoid antcin K by *Psychrobacillus* sp. Ak 187 [16]. Finally, Prof. Kovayashi reviews the total synthesis and biological evaluation of phaeosphaerides, where the use of catalysts can be appreciated for not only obtaining natural products but, also, their analogues for SAR studies and increasing the biological activity of synthesized compounds [17].

**In this manner, the reader, through this issue, could gain an idea of the new directions that the synthesis of natural products using catalysts will have in the years to come.**

**Conflicts of Interest:** The authors declare no conflict of interest.

## References

1.  Karageorgis, G.; Waldmann, H. Guided by Evolution: Biology-Oriented Synthesis of Bioactive Compound Classes. *Synthesis* **2019**, *51*, 55–66. [CrossRef]
2.  Gerry, G.J.; Schreiber, S.L. Recent achievements and current trajectories of diversity-oriented synthesis. *Curr. Opin. Chem. Biol.* **2020**, *56*, 1–9. [CrossRef] [PubMed]
3.  Wetzel, S.; Bon, R.S.; Kumar, K.; Waldmann, H. Biology-Oriented Synthesis. *Angew. Chem. Int. Ed.* **2011**, *50*, 10800–10826. [CrossRef] [PubMed]
4.  Wender, P.A.; Verma, V.A.; Paxton, T.J.; Pillow, T.H. Function-Oriented Synthesis, Step Economy, and Drug Design. *Acc. Chem. Res.* **2008**, *41*, 40–49. [CrossRef] [PubMed]
5.  Zhao, C.G.; Ye, Z.Q.; Ma, Z.X.; Scott, A.W.; Stephanie, A.B.; Hu, L.H.; Ilia, A.G.; Tang, W.P. A general strategy for diversifying complex natural products to polycyclic scaffolds with medium-sized rings. *Nat. Commun.* **2019**, *10*, 4015. [CrossRef] [PubMed]
6.  Special Issue on "Advances in Green Catalysis for Sustainable Organic Synthesis". Available online: https://www.mdpi.com/journal/catalysts/special_issues/Catalysis_Organic_Synthesis (accessed on 24 October 2019).
7.  Special Issue on "Sustainable and Environmental Catalysis of Catalysts". Available online: https://www.mdpi.com/journal/catalysts/special_issues/SEC_catalysts (accessed on 24 October 2019).
8.  Special Issue on "Biocatalysis and Biotransformations" of Catalysts. Available online: https://www.mdpi.com/journal/catalysts/special_issues/biocatal_trans (accessed on 24 October 2019).
9.  Special Issue on Combinatorial Chemistry of Molecules. Available online: https://www.mdpi.com/journal/molecules/special_issues/combinatorial-chemistry (accessed on 24 October 2019).
10. Special Issue on New Trends in Asymmetric Catalysis of Catalysts. Available online: https://www.mdpi.com/journal/catalysts/special_issues/Asymmet_Catal (accessed on 24 October 2019).
11. Special Issue on Organocatalysis: Advances, Opportunity, and Challenges of Catalysts. Available online: https://www.mdpi.com/journal/catalysts/special_issues/organocatal (accessed on 24 October 2019).
12. Chojnacka, A.; Gładkowski, W. Production of Structured Phosphatidylcholine with High Content of Myristic Acid by Lipase-Catalyzed Acidolysis and Interesterification. *Catalysts* **2018**, *8*, 281. [CrossRef]
13. Vila, C.; del Campo, A.; Blay, G.; Pedro, J.R. Catalytic Enantioselective Addition of $Me_2Zn$ to Isatins. *Catalysts* **2017**, *7*, 387. [CrossRef]
14. Pombal, S.; Tobal, I.E.; Roncero, A.M.; Rodilla, J.M.; Garrido, N.M.; Sanz, F.; Esteban, A.; Tostado, J.; Moro, R.F.; Sexmero, M.J.; et al. Diastereoselective Synthesis of 7, 8-Carvone Epoxides. *Catalysts* **2018**, *8*, 250. [CrossRef]
15. Ros Ñíguez, D.; Khazaeli, P.; Alonso, D.A.; Guillena, G. Deep Eutectic Mixtures as Reaction Media for the Enantioselective Organocatalyzed α-Amination of 1, 3-Dicarbonyl Compounds. *Catalysts* **2018**, *8*, 217. [CrossRef]
16. Chiang, C.-M.; Wang, T.-Y.; Ke, A.-N.; Chang, T.-S.; Wu, J.-Y. Biotransformation of Ergostane Triterpenoid Antcin K from *Antrodia cinnamomea* by Soil-Isolated *Psychrobacillus* sp. AK 1817. *Catalysts* **2017**, *7*, 299. [CrossRef]
17. Kobayashi, K.; Tanaka, K., III; Kogen, H. Total Synthesis and Biological Evaluation of Phaeosphaerides. *Catalysts* **2018**, *8*, 206. [CrossRef]

![catalysts logo] *catalysts*

MDPI

*Article*

# Production of Structured Phosphatidylcholine with High Content of Myristic Acid by Lipase-Catalyzed Acidolysis and Interesterification

**Anna Chojnacka \* and Witold Gładkowski**

Department of Chemistry, Wroclaw University of Environmental and Life Sciences, Norwida 25,
50-375 Wroclaw, Poland; glado@poczta.fm
* Correspondence: jeanna@wp.pl; Tel.: +48-661-464-355

Received: 6 June 2018; Accepted: 12 July 2018; Published: 14 July 2018

**Abstract:** Synthesis of structured phosphatidylcholine (PC) enriched with myristic acid (MA) was conducted by acidolysis and interesterification reactions using immobilized lipases as catalysts and two acyl donors: trimyristin (TMA) isolated from ground nutmeg, and myristic acid obtained by saponification of TMA. Screening experiments indicated that the most effective biocatalyst for interesterification was *Rhizomucor miehei* lipase (RML), whereas for acidolysis, the most active were *Thermomyces lanuginosus* lipase (TLL) and RML. The effect of the molar ratio of substrates (egg-yolk PC/acyl donor), enzyme loading, and different solvent on the incorporation of MA into PC and on PC recovery was studied. The maximal incorporation of MA (44 wt%) was achieved after 48 h of RML-catalyzed interesterification in hexane using substrates molar ratio (PC/trimyristin) 1/5 and 30% enzyme load. Comparable results were obtained in toluene with 1/3 substrates molar ratio. Interesterification of PC with trimyristin resulted in significantly higher MA incorporation than acidolysis with myristic acid, particularly in the reactions catalyzed by RML.

**Keywords:** immobilized lipases; structured phosphatidylcholine; myristic acid; trimyristin; acidolysis; interesterification; egg-yolk phosphatidylcholine

## 1. Introduction

Myristic acid (MA) is a 14-carbon lipid molecule belonging to long-chain saturated fatty acids. It usually accounts for small amounts (approximately 1 wt%) of total fatty acids (FA) in animal tissues, but is more abundant in milk fat (7–12 wt% of total FA) [1]. In copra and palmist oils, it makes up 15–23 wt% and 15–17 wt% of total FA, respectively [2], but the highest content of MA has been found in nutmeg (*Myristica fragrans*) in the form of trimyristin, which comprises approximately 45% by weight of the nutmeg butter [3].

Epidemiological and clinical studies have shown that dietary fats containing high levels of saturated fatty acids (SFA, usually more than 15% of total energy) induce an increase in plasma total cholesterol and low-density lipoprotein cholesterol (LDL cholesterol) concentrations in humans [4]. Along with other saturated FA, myristic acid has deleterious effects on blood cholesterol level in animals and humans when provided at high levels exceeding 4% of dietary energy [4,5], but these negative effects disappear when its dietary level is in a narrow physiological range (1.0–2.5% of dietary energy) [2,6]. In European countries, myristic acid consumption is 4–8 g per day, which corresponds to 0.5 to 2.0% of total energy. Furthermore, some studies have shown that specific biochemical functions can be assigned to MA; it is known to modify activity of many enzymes and protein functions of both eukaryotic and viral origin through the myristoylation of their N-terminal glycine residues [7–9]. The myristoyl residue of proteins increases their affinity to target cell membranes. Myristic acid may be also considered as one of the regulators of cellular bioactive lipid concentration, such as polyunsaturated fatty acids and ceramide [10]. The positive role of MA in the overall conversion of α-linolenic acid to Long Chain Polyunsaturated Fatty Acids (LC-PUFA), like eicosapentaenoic (EPA) and docosahexaenoic (DHA) acids, has been shown in rat and human nutritional experiments [2,11–13]. The effect of myristic acid on *n*-3 PUFA content follows a U-shaped curve, and a beneficial maximum level at around 1.2% of total daily energy is observed [13]. Positive correlation between myristic acid intake and the level of *n*-3 PUFA is connected with the myristoylation of NADH-cytochrome b5 reductase accounting for the increased Δ6-desaturase activity. In turn, the role of MA in the biosynthesis of ceramide lies in the myristoylation of dihydroceramide Δ4-desaturase. MA also exerted anxiolytic-like effects, producing comparable actions to diazepam in Wistar rats subjected to the elevated plus maze [14].

Phospholipids (PLs) are widely used in the food, cosmetic, and pharmaceutical industries [15]. Interest in the production of structured phospholipids (SPL) containing specific fatty acid residues has grown significantly in recent years. Replacement of existing fatty acids in native PL with desirable fatty acids can improve not only their physical and chemical properties, but also their nutritional, pharmaceutical, and medical functions [16]. Novel SPL were created by incorporating into natural PL functional fatty acids that have multiple biochemical and pharmacological effects on human health and treatment/prevention of some diseases (e.g., docosahexaenoic acid (DHA), eicosapentaenoic acid (EPA), and conjugated fatty acids) [17–24].

Compared with chemical methods, enzymatic modification of PL has advantages that are attributed to the specificity of the enzymes. Lipases of microbial origin and phospholipase $A_1$, namely Lecitase Ultra, have been the most commonly used enzymes for the exchange of fatty acids on PL at *sn*-1 positions, whereas phospholipase $A_2$ have been employed to the modification of *sn*-2 position [15,17,19,20,25,26]. Regioselectivity of lipases allows the specific removal or/and replacement of the acyl chains at position *sn*-1 of PL via hydrolysis followed by re-esterification or through direct transesterification with acyl donors (acidolysis with free fatty acids and interesterification with esters of fatty acids).

In our previous study, we reported on a pioneer method for the production of structured phosphatidylcholine (SPC) using plant oils as the source of natural acyl donors in the form of triacylglycerols. Using this methodology, phospholipids were enriched with unsaturated fatty acids from the *n*-3 or *n*-6 family in the process of interesterification with corresponding oil, that is, linseed, safflower, sunflower, borage, and primrose evening oil [27]. The same reaction system with pomegranate seed oil as the acyl donor was applied in the incorporation of conjugated linolenic acid (punicic acid) into egg-yolk phosphatidylcholine (PC) [21]. In turn, concentrates of *n*-3 PUFA and CLA (conjugated linolenic acid) produced from fish oil and sunflower oil, respectively, were applied to the enrichment of egg-yolk PC with desired fatty acids in the process of acidolysis [18,22].

Phospholipids (PLs) can be also an efficient dietary vehicle to deliver MA into the target cells for the myristoylation of proteins. Werbovetz and Englund reported 1-myristoyllysophosphatidylcholine (M-LPC) as myristate donor for the myristoylation of glycosyl phosphatidylinositol (GPI) anchor of the *Trypanosoma brucei* variant surface glycoprotein (VSG) [28]. M-LPC is first hydrolyzed by plasma membrane-associated phospholipase to release a cell-associated MA that, after uptake by the cell and conversion to mirystoyl-CoA, becomes available for the myristoylation of GPI catalyzed by myristoyltransferase.

Considering the role of myristic acid in the proteins activation and the positive effect of myristic acid on the EPA and DHA levels when supplied at around 1.2% of daily energy, we were encouraged to produce new PC species containing MA in the sn-1 position. As M-LPC, similar to other LPC species, can be easily accessible from 1-myristoyl-2-acyl-PC via the hydrolysis catalyzed by phospholipase $A_2$ [29], we decided to elaborate the novel methods for incorporation of MA into the *sn*-1 position of egg-yolk PC. In the presented work, continuing our studies on the enzymatic production of PLs using acyl donors of natural origin, we report on the lipase catalyzed-production of MA-enriched PC using two acyl donors: trimyristin (TMA) prepared from nutmeg, and MA obtained from trimyristin by saponification. Using these acyl donors allowed us to compare two reaction systems: interesterification and acidolysis.

## 2. Results and Discussion

The lipase-catalyzed reaction systems that were applied to obtain MA-riched PC are depicted in Scheme 1.

The raw material for our studies were PC isolated from egg-yolk (purity >99% according to HPLC) and two acyl donors (trimyristin and myristic acid). TMA was isolated from ground nutmeg by continuous Soxhlet extraction with hexane and subsequent crystallization from acetone. Myristic acid was obtained by saponification of TMA isolated previously from nutmeg.

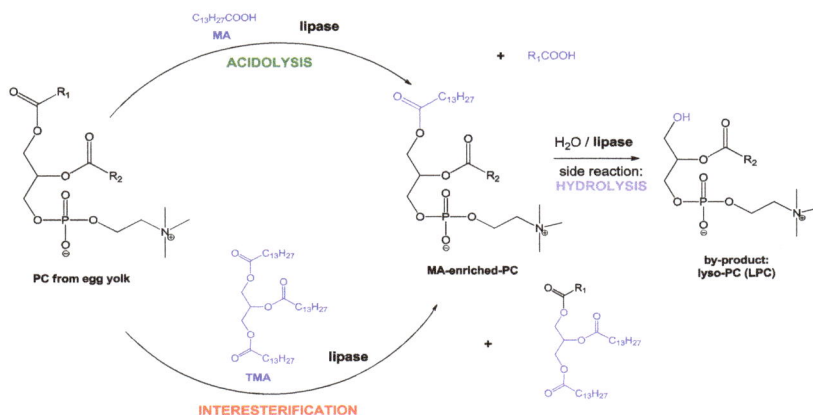

**Scheme 1.** Lipase-catalyzed incorporation of myristic acid into egg-yolk phosphatidylcholine (PC) with partial hydrolysis of modified PC as a side reaction. MA—myristic acid; TMA—trimyristin; LPC—lysophosphatidylcholine.

PC, TMA, and MA were subjected to gas chromatography (GC) analysis to determine their fatty acid composition (Table 1). The results for PC showed that palmitic (C16:0) and oleic (C18:1) acids were the predominating ones, and the former, as well as stearic acid (C18:0), are located in the *sn*-1 position, whereas the *sn*-2 position is occupied by unsaturated acids from the *n*-6 (linoleic acid and arachidonic acid) and *n*-9 family (oleic acid).

**Table 1.** Fatty acid (FA) composition (wt%) of substrates used in lipase-catalyzed transesterifications of egg-yolk phosphatidylcholine (PC) with trimyristin (TMA) and myristic acid (MA).

| FA | TMA | MA | Native PC | | | Modified PC [2] | | |
|---|---|---|---|---|---|---|---|---|
| | | | Total | *sn*-1 | *sn*-2 | Total | *sn*-1 | *sn*-2 |
| C14:0 | 83.2 ± 0.21 [1] | 92.3 ± 0.31 | - | - | - | 44.0 ± 0.14 | 80.5 ± 0.25 | 7.1 ± 0.11 |
| C16:0 | 6.1 ± 0.45 | 6.2 ± 0.11 | 33.2 ± 0.11 | 64.1 ± 0.22 | 4.4 ± 0.04 | 8.3 ± 0.05 | 12.1 ± 0.11 | 4.3 ± 0.07 |
| C16:1 | - | - | 1.2 ± 0.05 | 1.1 ± 0.12 | 1.5 ± 0.05 | 1.1 ± 0.03 | 1.0 ± 0.07 | 1.7 ± 0.12 |
| C18:0 | - | - | 15.7 ± 0.05 | 29.2 ± 0.55 | 2.7 ± 0.09 | 2.7 ± 0.11 | 2.5 ± 0.05 | 2.7 ± 0.15 |
| C18:1 | 5.3 ± 0.05 | 1.1 ± 0.02 | 29.0 ± 0.04 | 4.6 ± 0.33 | 57.1 ± 0.32 | 23.9 ± 0.04 | 2.6 ± 0.12 | 45.5 ± 0.44 |
| C18:2 | - | - | 15.3 ± 0.02 | 1.0 ± 0.05 | 29.8 ± 0.55 | 14.8 ± 0.21 | 1.0 ± 0.04 | 29.3 ± 0.55 |
| C20:4 | - | - | 3.0 ± 0.01 | - | 4.5 ± 0.06 | 5.2 ± 0.22 | 0.3 ± 0.05 | 9.4 ± 0.05 |
| unidentified | 5.4 ± 0.08 | 0.4 ± 0.03 | - | - | - | - | - | - |

[1] Data are presented as mean ± SD of two independent analysis; [2] Reaction conditions: enzyme *Rhizomucor miehei* lipase (RML)—load 30% (*w*/*w*), 1/5 PC/TMA molar ratio, temperature 50–52 °C, hexane, 48 h.

Analysis of the FA composition of trimyristin indicated the high content of myristic acid (over 83 wt%) and some small amounts of other acids (about 6 wt%, C16:0 and 5 wt%, C18:1). The purity of myristic acid obtained by saponification of TMA, given as wt% of myristic acid according to GC, was approximately 92%.

## 2.1. Screening of Enzymes

Four commercially available lipases; Lipozyme® (RML), Lipozyme TL IM (TLL), *Candida antarctica* lipase B (Novozym 435) and *Candida antarctica* lipase A (CALA) were evaluated for their ability to catalyze interesterification and acidolysis between egg-yolk PC and TMA or MA, respectively. Lipozyme® and Lipozyme TL IM are classified as 1,3-regiospecific lipases [15] the third one is considered as a non-specific enzyme, but in most lipid modifications it shows high selectivity towards the *sn*-1 position [27,30,31]. The fourth of the lipases (CALA) is reported to show a *sn*-2-regioselectivity [32,33], although such preference in the structure of triacylglycerols (TAG) is not sufficient to selective synthesis of 1,3-diglycerides or 2-monoglycerides [34]. Therefore, for practical interesterifications, CALA has been reported rather as a nonselective lipase [35]. All enzymes were used in immobilized form, which made it possible to carry out the reactions in the presence of solvent and at higher temperatures.

The initial conditions applied for both reaction systems (acidolysis and interesterification) were as follows: temperature 50–52 °C; 1/3 PC/acyl donor molar ratio and 30% enzyme dosage. Having in mind a decrease of process costs, the employed lipases were used at the same weight ratios, despite different enzymatic activities declared by the suppliers.

The time course of the incorporation of MA into PC by lipases in both processes is shown in Figure 1.

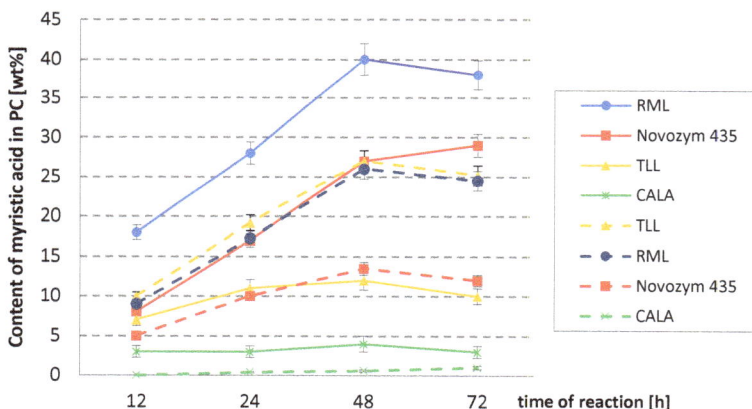

**Figure 1.** Time courses of different lipases-catalyzed acidolysis (- - -) and interesterification (—) between egg-yolk phosphatidylcholine (PC) and myristic acid or trimyristin. Reaction condition: temperature, 50–52 °C; PC/acyl donor molar ratio, 1/3; lipase dosage, 30%; solvent, hexane. RML—*Rhizomucor miehei* lipase; TLL—*Thermomyces lanuginosus* lipase; CALA—*Candida Antarctica*.

The screening experiments indicated that immobilized lipases from *Rhizomucor miehei* (Lipozyme RM IM, RML) and *Thermomyces lanuginosus* (Lipozyme TL IM, TLL) exhibited the highest activity. The first one was the most efficient in interesterification (40 wt% incorporation of MA into egg-yolk PC after 48h), whereas in acidolysis, both mentioned lipases exhibited satisfactory activity. However, less incorporation degree was observed at the same time. In the reaction catalyzed by TLL, the incorporation of MA reached a maximum (27 wt%) within 48 h, whereas in RLM-catalyzed acidolysis, 26 wt% of incorporation was observed. In this investigation, Novozym 435 showed less activity, giving 27.5 wt% and 13.5 wt% incorporation of MA into PC in interesterification and acidolysis, respectively. In our previous studies, this lipase was the most efficient in incorporation of α-linolenic acid, punicic acid, and PUFA into PC in the interesterification and acidolysis processes [18,21,27,30]. On the other hand, similar to the results presented here, TLL was less active in interesterification of PC with pomegranate seed oil. Comparing acidolysis of PC with concentrates from fish oil and myristic acid, higher selectivity of TLL towards MA was observed.

Lipase A from *Candida antarctica* was almost inactive in both the interesterification and acidolysis processes; after 72 h of reactions, no more than 4 wt% incorporation of MA was achieved.

Collating efficiency of MA incorporation, about two-fold higher introduction of myristic acid in the process of interesterification than in acidolysis catalyzed by both RML and Novozym 435 can be seen. The reason is the fact, that compared with myristic acid, trimyristin is a more efficient acyl donor, containing two acyl groups in the *sn*-1 and *sn*-3 positions, both accessible in interesterification of PC for 1,3-regioselective lipases used as biocatalysts.

Because the highest degree of incorporation of MA into PC in interesterification was achieved for RML, while in acidolysis, comparably high results were achieved for TLL and RML, these enzymes were selected for subsequent experiments.

## 2.2. Effect of Substrate Molar Ratio

One of the most important factors affecting both esterification and transesterification reactions is an acyl donor concentration [36].

The effect of substrates molar ratio (1/1, 1/3, and 1/5; PC/acyl donor: MA or TMA) on the degree of myristic acid incorporation into PC was evaluated for RML lipase in both interesterification and acidolysis and for TLL in the acidolysis process (Figure 2).

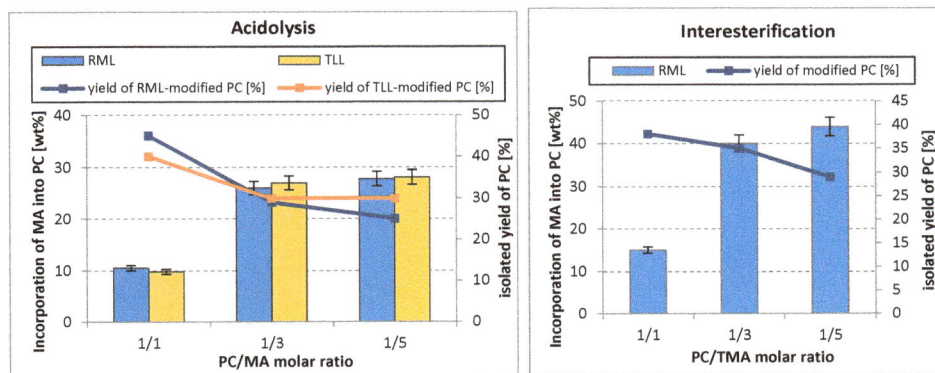

**Figure 2.** Effect of substrate molar ratio on the incorporation of myristic acid into egg-yolk PC and isolated yield of modified PC in acidolysis and interesterification (reaction condition: temperature, 50–52 °C; lipase dosage, 30%; solvent, hexane; 48 h). MA—myristic acid; TMA—trimyristin.

The results showed that for both lipases and both compared processes, the content of myristic acid in modified PC rose with an increasing substrates molar ratio with a simultaneous decrease in isolated yield of PC. In both reaction systems, the highest increase of MA incorporation into PC was observed when the ratio changed from 1/1 to 1/3 (PC/acyl donor). An almost three-fold growth of incorporation (from 9.8 to 27 wt%) can be observed for TLL-catalyzed acidolysis. In the case of the RML, increase of MA incorporation was significantly higher in the process of interesterification (from 15 to 40 wt%) than in acidolysis (from 10.5 to 26 wt%). A further enhancement of PC/acyl donor to 1/5 molar ratio caused an insignificant rise of embedded acid, whose content in modified PC made up about 28 wt% of total FA after acidolysis and 44 wt% after interesterification. It should be noticed that in both reaction systems, equilibrium is moved to the product side with an increasing substrate ratio, which improves acyl incorporation. Simultaneously, the overall PC yield decreased, which may be the result of hydrolysis side-reaction with 2-acyl LPC formation (Scheme 1) and next spontaneous acyl migration resulting in 1-acyl LPC, which may be quickly hydrolyzed to glycerophosphocholine (GPC). A drop of isolated yield of modified PC may be also related with difficulties in separating the unreacted acyl donors and exchanged fatty acids or triacylglycerols from the final products. As a result of that and taking into consideration the economy of the process, 1/3 molar ratio (PC/acyl donor) was chosen for subsequent experiments.

The effect of the substrate molar ratio on the incorporation of other saturated FA into phosholipids was investigated in several studies. Reddy et al. optimized the molar ratio PC/palmitic acid, changing it from 1:2 to 1:10, and established 1/5 molar ratio for maximal incorporation of palmitic acid into PC in the reaction catalyzed by Lipozyme TL IM [37]. Further increase of this parameter from 1/5 to 1/10 did not change the incorporation significantly. During acidolysis of PL with caprylic acid catalyzed by TLL in a packed bed reactor, the increase of incorporation with increasing substrate molar ratio from 1/4 to 1/36 PL/caprylic acid was observed, reaching 35.8% of incorporation [38]. On the other hand, in the solvent-free system, the incorporation of caprylic acid was 13.1 and 24.4% with PC/caprylic acid ratios of 1/6 and 1/8, respectively.

### 2.3. Effect of Enzyme Dosage

Enzyme dosage was the most significant factor on the lipase-catalyzed acidolysis and interesterification in terms of incorporation of MA into PC and PC recovery.

Incorporation into PC increased with increasing enzyme dosage (Figure 3). The most meaningful rise of MA in PC, over nine-fold for TLL and three-fold for RML-catalyzed acidolysis, was observed when the enzyme load was increased from 10 up to 20 wt%. Simultaneously, a decrease of modified PC yield was appreciable (reduced from 47 to 33 wt% for RML and from 55 to 35 wt% for TLL). A similar effect can be noticed for interesterification process catalyzed with RML. Incorporation of MA raised from 15 to 29 wt%, while the total isolated yield of PC decreased over 1.5-fold. It has been reported earlier that high enzyme dosage is important to achieve effective incorporation of fatty acids into PLs by acidolysis [33,34], but with the increase of the enzyme concentration, usually hydrolysis of PC proceeded lowering the total isolated yield of PC [35].

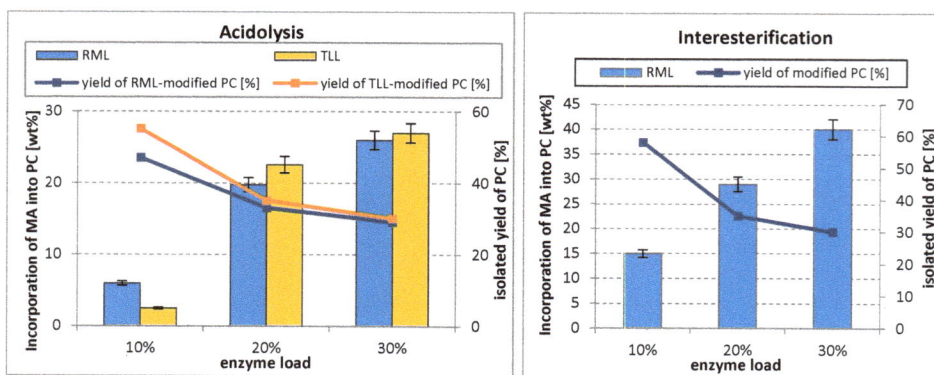

**Figure 3.** Effect of enzyme dosage on the incorporation of myristic acid into egg-yolk PC and isolated yield of modified PC in acidolysis and interesterification (reaction condition: temperature, 50–52 °C; PC/acyl donor molar ratio, 1/3; solvent, hexane; 48 h).

Increasing the dosage of the enzyme to 30% in the case of acydolysis resulted in a slight increase of myristic acid in PC (26 and 27 wt% incorporation for RML and TLL, respectively), with a little decrease of the total yield of modified PC. A clear increase of MA incorporation (from 29 up to 40 wt%) was observed for RML-catalyzed interesterification, as well as a small reduction of PC recovery.

In summary, with the increase of the enzyme dosage in both reaction systems, higher incorporation of MA can be obtained with simultaneous losses of PC. However, taking into consideration the cost of the enzyme, 20 wt% of enzyme load seems to be enough to achieve high incorporation of MA with a sufficient PC yield. The same enzyme dosage was found to be optimal by Reddy et al., who studied the effect of lipase dosage on the incorporation of palmitic and stearic acid in the TLL-catalyzed acidolysis of PC from egg-yolk [37]. Increasing enzyme dosage from 5 to 20%, the incorporation of palmitic acid and stearic acid reached up to 21.2% and 36.1%, respectively. Further increase of the enzyme load to 25% did not affect the incorporation of both acids.

### 2.4. Effect of Organic Solvents

Three organic solvents were tested in both reaction systems. Generally, no significant differences in MA incorporation were observed in the case of interesterification, the content of myristic acid in PC was in the range of 40–45 wt% for all solvents used. In the case of acidolysis, the effect of the solvent was pronounced for both RML and TLL (Figure 4). The highest incorporation (about 26 wt%) was obtained after acidolysis was carried out in hexane. Using toluene resulted in lowering incorporation by 6 wt% and a relevant drop in the content of MA in modified PC (to 10 wt% of total fatty acid) was observed for diisopropyl ether (DIPE), particularly for TLL-catalyzed reaction. In both reaction systems, increase of PC yield was found when DIPE was applied.

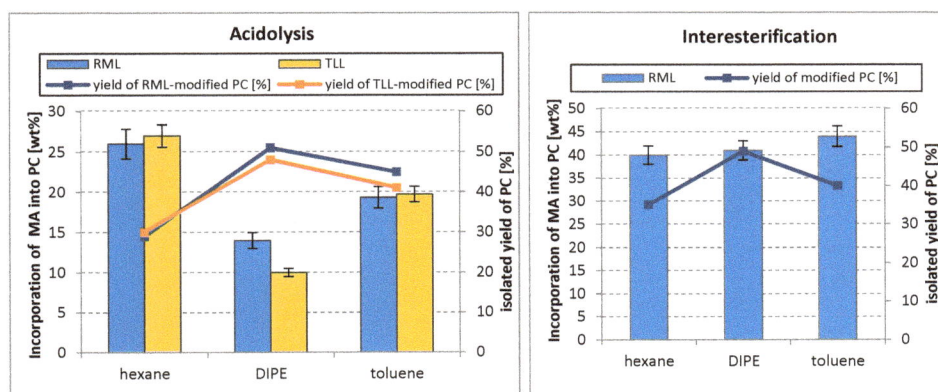

**Figure 4.** Effect of organic solvent on the incorporation of myristic acid into egg-yolk PC and isolated yield of modified PC in acidolysis and interesterification (reaction condition: temperature, 50–52 °C; PC/acyl donor molar ratio, 1/3; lipase dosage, 30%; 48 h). DIPE—diisopropyl ether.

Polarity of the solvent is crucial for holding a layer of water around the enzyme, and for this purpose, hydrophobic solvents are the best choice. Logarithm of the partition coefficients (log P) for the solvents used were as follows: 3.5 for hexane, 2.5 for toluene, and 1.52 for diisopropyl ether (DIPE). Solvents with log P in the range of 2–4 are considered as suitable for lipase-catalyzed reactions and the results of acidolysis confirmed the relationship between solvent polarity and enzyme activity. However, using DIPE in the interesterification gave MA incorporation values comparable to those obtained for hexane and toluene. No effect of solvent on the incorporation of caprylic acid into PC was also observed by Kim and Yoon during acidolysis catalyzed by *Mucor javanicus* lipase [39]. They used three solvents with significantly different polarity indexes: hexane, diethyl ether, and methanol; but in all cases, incorporation of caprylic acid was approximately 36%.

There have been several reports on the introduction of saturated fatty acids into phospholipids using enzymes to improve emulsifying and dispersing properties or heat stability, but only few concern myristic acid. Gan et al. introduced a series of saturated fatty acids (C6:0, C8:0, C10:0, C12:0, C14:0) into the *sn*-1 position of lecithin (mixture of phospholipids with ≥30% of PC) by Lecitase Ultra-catalyzed acidolysis (showing both lipase and phospholipase A$_1$ activity) [40]. They achieved different molar incorporation values, depending on the chain length; the highest one was observed for capric acid (C10:0, 51%), whereas the lowest was for caproic acid (C6:0, 28.9%). The reactions were carried out in hexane at high substrates molar ratio (10/1 MA/lecithin) and 10% of immobilized enzyme load and incorporation of myristic acid in these conditions was 36.4%. Caprylic acid (C8:0) was also successfully introduced into PC by different lipases: RML, TLL, and *Mucor javanicus* lipase; incorporation values were 46.3%, 35.4%, and 36.7%, respectively [38,39,41]. TLL was also used for the acidolysis of soybean phospholipids with capric acid (C10:0) and 23.16% of incorporation was achieved after 10 days of reaction at hexane, at 1/4 substrates molar ratio (PL/capric acid) and 10% enzyme load [42]. An interesting method was proposed by Hama et al. [43]. Using immobilized *Rhizopus oryzae* whole-cells at 30 °C in hexane at 1/8 substrates molar ratio (PC/fatty acid), introduction of lauric acid (C12:0) into egg-yolk PC at the level of 44.2% was achieved. The biocatalyst used was more efficient than different lipase powders. Lipase-catalyzed enrichment of PC with palmitic (C16:0) and stearic acid (C18:0) was performed by Reddy et al.; the best results for egg-yolk PC were obtained using TLL when the content of 16:0 was increased from 37.4% to 58.6%, and stearic acid from 8.6% to 44.7%, whereas for soybean PC, higher incorporation (44.2% for 16:0 and 53.7% for C 18:0) was found for Novozyme 435. In these studies, 20% enzyme load, 1/5 PC/FA molar ratio, and reaction time 24 h were chosen as optimal conditions for maximal incorporation of both incorporated acids into PC [37].

Egger et al. [44] studied the phospholipase A$_2$-catalyzed esterification of LPC with different fatty acids, including myristic acid. The reaction carried out in toluene afforded 1-palmitoyl-2-myristoyl-PC, but the reaction rate was low (below 2 nmol of product per h and mg of enzyme preparation) and no information about the PC yield is given. The only one report concerns the introduction of myristic acid into the *sn*-1 position of PLs using lipase is the interesterification between soybean PC and methyl myristate catalyzed by RML [45]. Using a solvent free system and 1/5 substrate molar ratio (soy PC/methyl myristate), as well as 10% of the enzyme load, the authors achieved 15.7% MA incorporation in the reaction conducted at 60 °C. Using methyl caprate or methyl laurate as the acyl donors, authors achieved 8.4% and 14.1% of incorporation of capric acid and lauric acid, respectively. No further studies on the effect of reaction condition were presented.

In this paper, for the first time, the complex studies on the lipase-catalyzed incorporation of myristic acid from natural source into egg-yolk PC were presented. Similar to most other investigations concerning incorporation of saturated fatty acids into phospholipids, RML and TLL turned out to be the most active biocatalysts. Comparison of two reaction systems proved higher efficiency of interesterification with trimyristin when compared with acidolysis with myristic acid. The use of 1,3 regioselective lipases allowed as to produce PC enriched with MA mainly in the *sn*-1 position (over 80 wt%), leaving unsaturated acids (45.5 wt% C18:1, 29.3 wt% C18:2, and 9.4 wt% C20:4) in their original, internal position (Table 1). The results show the characteristic decrease of palmitic and stearic acid (from 64.1 to 12.1 wt% and from 29.2 to 2.5 wt%, respectively) in the external position of PC because of their replacement by myristic acid. These results confirm the regioselectivity of lipases used, a small amount of myristic acid in the *sn*-2 position may be the result of acyl migration during the processes.

An elaborated method of interesterification with *Rhizomucor miehei* lipase and trimyristin as the acyl donor led to significantly higher incorporation degree (40–44 wt%) of myristic acid than those reported by Ghosh [45], at the lower substrate molar ratio (1/3, PC/acyl donor) and lower temperature

(50–52 °C). The presented results confirm the usefulness of natural triacylglycerols as the effective acyl donors in the interesterifications of phospholipids.

## 3. Materials and Methods

### 3.1. Materials and Chemicals

Lohmann Brown hens' eggs were purchased from the poultry farm "Ovopol" (Nowa Sól, Poland). Lipozyme TL IM (a silica granulated *Thermomyces lanuginosus* lipase preparation, 250 U/g) was a gift from the Novozymes A/S (Bagsvaerd, Denmark). Lipase B from *Candida antarctica* immobilized in a macroporous acrylic resin (synonym: Novozym 435, >5000 U/g), lipase A from *Candida antarctica* (CALA, >500 U/g) immobilized on resin Immobead 150, and lipase from *Mucor miehei* immobilized in macroporous ion-exchange resin (Lipozyme®, >30 U/g) were purchased from Sigma-Aldrich (St. Louis, MO, USA). A boron trifluoride methanol complex solution (13–15% BF$_3$ × MeOH) was purchased from Sigma-Aldrich (St. Louis, MO, USA). All other chemicals were of analytical grade. Silica gel-coated aluminum plates (Kieselgel 60 F254, 0.2 mm) used in thin layer chromatography (TLC) and the silica gel (Kieselgel 60, 230–400 mesh) used in the column chromatography were purchased from Merck.

### 3.2. Analysis of Substrates and Products

The purity of the native and modified phosphatidylcholine was determined by HPLC on an Ultimate 3000 DIONEX chromatograph equipped with Corona™ Charged Aerosol Detector (CAD). A Waters Spherisorb S5W column (150 × 4.6 mm) was used for analysis. The HPLC gradient program was as follows: (flow rate 0.6 mL × min$^{-1}$); 0 min 0/90/10 (%A/%B/%C) at 2 min, 0/40/60 at 20 min, 1/40/59 at 22 min, 10/40/50 at 38 min, 8/40/52 at 44 min, 1/40/59 at 55 min, 0/90/10 at 56 min, and 0/90/10 hold 10 min (A/B/C, water/0.1% solution of formic acid in hexane/isopropanol).

Fatty acid profiles of starting materials and structured PC were determined after their conversion to the fatty acid methyl esters (FAME) according to the following procedure: samples (50 mg) were heating under reflux (3 min) with 3 mL of BF$_3$ × MeOH complex solution. After cooling, the mixtures were extracted with 2 mL of hexane and the organic layers were washed with a saturated NaCl solution. Hexane extracts were dried over anhydrous magnesium sulphate (MgSO$_4$) and analyzed directly by gas chromatography (GC) on an Agilent 6890 N instrument using a 70% cyanopropyl polysilphenylene-siloxane column (TR FAME, 30 m × 0.25 mm × 0.25 µm) and flame ionization detector. The temperature parameters were as follows: injector 250 °C, detector 280 °C, column: 160 °C (held 3 min), 160–220 °C (rate 5 °C·min$^{-1}$), 220–260 °C (rate 30 °C·min$^{-1}$), 260 °C (held 3 min). The FAME were identified by comparing their retention times with those of a standard FAME mixture (Supelco 37 FAME Mix) purchased from Sigma-Aldrich.

The incorporation of myristic acid into PC is refered to the total amount of PC species obtained after enzymatic modification (original and modified PC molecules).

The positional analysis of fatty acids in native and modified PC was based on regiospecific Lipozyme®-catalyzed ethanolysis of PC, which allowed us to obtain fatty acids ethyl esters (FAEE) released from the *sn*-1 position of PC and 2-acyl LPC. FAEE were analyzed directly by gas chromatography to give the composition of FA in the *sn*-1 position of PC. After purification, 2-acyl LPC were trans-esterified using ethanolic NaOH solution and boron trifluoride etherate. FAEEs obtained were isolated from the reaction mixture and analyzed by gas chromatography. In this way, FAs composition in the *sn*-2 position of the PC was determined. The details of the procedure were described in our previous paper [46].

### 3.3. Isolation of PC from Egg Yolk

The extraction of phospholipids from egg yolk was performed on a semi-technical scale in Wroclaw Technology Park. Eggs were dried in the drying chamber at inlet air temperature 185 ± 5 °C and an outlet air temperature 70 ± 2 °C. In the next step, obtained powder was extracted with ethanol in a tank equipped with a mechanical stir maintaining the yolk/solvent ratio 1:4 (*m*/*v*). The process of suspension was carried out for 90 min and then alcohol was removed by filtration. The residue was evaporated in vacuo (0.06 MPa at 50 °C). The pure PL-fraction was obtained from the crude PL by precipitation with cold acetone [47]. The pure PC was separated from this fraction by silica gel column chromatography (chloroform/methanol/water, 65:25:4, *v/v/v*). The purity of PC fractions was analyzed by TLC on silica gel-coated aluminum plates (chloroform/methanol/water, 65:25:4, *v/v/v*) and HPLC. Fractions containing more than 99% of PC (according to HPLC) were collected and the solvent was evaporated.

### 3.4. Preparation of Acyl Donors

#### 3.4.1. Isolation of Trimyristin from Nutmeg

Finely ground nutmeg (25 g) was placed into the extraction thimble, covered with a little cotton wad, and placed in the Soxhlet extractor. Hexane (200 mL) was poured into the flask and heated to reflux. The extraction was continued for 2–3 h (10–15 extraction cycles). The solvent was evaporated from the extract and the residue was crystallized from acetone. After 2 h, crystals of trimyristin were filtered and dried in the desiccator with anhydrous calcium chloride. The fatty acid compositions [wt%] of trimyristin were analyzed by gas chromatography (Table 1).

#### 3.4.2. Production of Myristic Acid from Trimyristin by Saponification

The mixture of TMA isolated previously from nutmeg (10 g) dissolved in 150 mL of ethanol and 100 mL 1 M NaOH ethanolic solution was heated (78 °C) under reflux for 3 h. Then, the mixture was cooled to room temperature, poured into 100 mL of water in a large beaker, and acidified with 20 mL of concentrated hydrochloric acid to pH = 2. After 30 min, white crystals of myristic acid were filtrated on a Büchner funnel, washed with 25 mL of distilled water, and dried in laboratory drier at 40 °C overnight to afford 8.7 g of myristic acid (yield 92%).

The progress of the saponification reaction was monitored by thin layer chromatography (TLC) on a silica gel-coated aluminum plates (developing system—hexane/diethyl ether (3/1, *v/v*)). After elution, the plates were developed using the 0.05% primuline solution (acetone/ water, 8:2, *v/v*) and spots were detected under an ultraviolet (UV) lamp (λ = 320 nm).

The purity of myristic acid [wt%] was determined by GC (Table 1).

### 3.5. The Lipase-Catalyzed Acidolysis of PC with Myristic Acid

The egg-yolk PC (0.325 mmol, 200 mg) and myristic acid (at molar ratio of substrates 1/3, PC/MA) in 5 mL of solvent was mixed at 50 °C until complete dissolution and then 20% of lipase (by weight of substrates) was added. The reactions were carried out using four different lipases at 50–52 °C. The effect of molar ratio of substrates, lipase dosage, and different solvent was tested in another set of experiments for two enzymes with high activity: Lipozyme® and Lipozyme TL IM. Each experiment was carried out in two replications. The reaction mixtures were agitated in a heating plate with a magnetic stirrer at 300 rpm, and stopped at the selected time intervals by enzyme filtration.

After evaporation of solvent in vacuo PC was separated from the other components by column chromatography using silica gel as stationary phase and the mixture of chloroform/methanol/water (65:25:4, *v/v/v*) as mobile phase (eluent). The procedure was as follows: silica gel was well solvated with chloroform and the prepared slurry was gently poured into the glass column. The sample was dissolved in a minimal (2 mL) amount of chloroform and applied to the top of the column and then eluted with the solvent system mentioned above. The eluted fractions were collected in test tubes and identified by TLC on silica gel-coated aluminum plates (chloroform/methanol/water, 65:25:4, *v/v/v*) and HPLC. Fractions of PC were combined and dried with anhydrous magnesium (VI) sulfate. After filtration, the solvent was evaporated *in vacuo* to obtain pure PC (>99% according to HPLC)

Isolated yield was calculated as the total amount (by weight) of PC recovered after the process (which contains modified and unmodified PC) in relation to initial PC.

*3.6. The Lipase-Catalyzed Interesterification of PC with Trimyristin*

The egg-yolk PC (0.325 mmol, 200 mg) was mixed with TMA (at 1/3 molar ratio of substrates, PC/TMA) in 5 mL of hexane and after dissolving of substrates, lipase (20% by weight of substrates) was added. The reactions were carried out using four different lipases, at 50–52 °C. The effect of molar ratio of substrates, lipase dosage, and different solvent was tested in another set of experiments for Lipozyme®. Each experiment was performed in two replications. The reaction mixtures were agitated in a heating plate with a magnetic stirrer at 300 rpm, and stopped at the selected time intervals by enzyme filtration. Modified PC was separated from the mixtures by silica-gel column chromatography (chloroform/methanol/water, 65:25:4, *v*/*v*/*v*) according to the same procedure described in Section 3.5.

Isolated yield was estimated as the total amount (by weight) of PC recovered after the process (which contains modified and unmodified PC) in relation to initial PC.

**Author Contributions:** A.C. conceived and designed the experiments; A.C. and W.G performed the experiments, gas chromatography analysis, and HPLC analysis; A.C. analyzed the data and wrote the paper.

**Funding:** This research received no external funding.

**Acknowledgments:** Publication supported by Wroclaw Centre of Biotechnology, Programme The Leading National Research Centre (KNOW) for years 2014–2018 (http://know.wroc.pl).

**Conflicts of Interest:** The authors declare no conflict of interest.

## References

1. Jensen, R.G.; Ferris, A.M.; Lammi-Keefe, C.J.; Henderson, R.A. Lipids of Bovine and Human Milks: A Comparison. *J. Dairy Sci.* **1990**, *73*, 223–240. [CrossRef]
2. Rioux, V.; Catheline, D.; Bouriel, M.; Legrand, P. Dietary myristic acid at physiologically relevant levels increases the tissue content of C20:5 *n*-3 and C20:3 *n*-6 in the rat. *Reprod. Nutr. Dev.* **2005**, *45*, 599–612. [CrossRef] [PubMed]
3. Power, F.B.; Salway, A.H. CLXII.—The constituents of the expressed oil of nutmeg. *J. Chem. Soc. Trans.* **1908**, *93*, 1653–1659. [CrossRef]
4. Salter, A.; Mangiapane, E.; Bennett, A.; Bruce, J.; Billett, M.; Anderton, K.; Marenah, C.; Lawson, N.; White, D. The effect of different dietary fatty acids on lipoprotein metabolism: Concentration-dependent effects of diets enriched in oleic, myristic, palmitic and stearic acids. *Br. J. Nutr.* **1998**, *79*, 195–202. [CrossRef] [PubMed]
5. Hayes, K.C.; Khosla, P. Dietary fatty acid thresholds and cholesterolemia. *FASEB J.* **1992**, *6*, 2600–2607. [CrossRef] [PubMed]
6. Dabadie, H.; Motta, C.; Peuchant, E.; LeRuyet, P.; Mendy, F. Variations in daily intakes of myristic and α-linolenic acids in *sn*-2 position modify lipid profile and red blood cell membrane fluidity. *Br. J. Nutr.* **2006**, *96*, 283–289. [CrossRef] [PubMed]
7. Towler, D.A.; Gordon, J.I.; Adams, S.P.; Glaser, L. The biology and enzymology of eukaryotic protein acylation. *Annu. Rev. Biochem.* **1988**, *57*, 69–99. [CrossRef] [PubMed]
8. Johnson, D.R.; Bhatnagar, R.S.; Knoll, L.J.; Gordon, J.I. Genetic and Biochemical Studies of Protein N-Myristoylation. *Annu. Rev. Biochem.* **1994**, *63*, 869–914. [CrossRef] [PubMed]
9. Beauchamp, E.; Rioux, V.; Legrand, P. Acide myristique: Nouvelles fonctions de régulation et de signalisation. *Medecine/Sciences* **2009**, *25*, 57–63. [CrossRef] [PubMed]
10. Rioux, V. Fatty acid acylation of proteins: Specific roles for palmitic, myristic and caprylic acids. *Oilseeds Fats Crop. Lipids* **2016**, *23*, D304. [CrossRef]
11. Legrand, P.; Beauchamp, E.; Catheline, D.; Pédrono, F.; Rioux, V. Short chain saturated fatty acids decrease circulating cholesterol and increase tissue pufa content in the rat. *Lipids* **2010**, *45*, 975–986. [CrossRef] [PubMed]
12. Jan, S.; Guillou, H.; D'Andrea, S.; Daval, S.; Bouriel, M.; Rioux, V.; Legrand, P. Myristic acid increases Δ6-desaturase activity in cultured rat hepatocytes. *Reprod. Nutr. Dev.* **2004**, *44*, 131–140. [CrossRef] [PubMed]

13. Legrand, P.; Rioux, V. The complex and important cellular and metabolic functions of saturated fatty acids. *Lipids* **2010**, *45*, 941–946. [CrossRef] [PubMed]

14. Contreras, C.M.; Rodríguez-Landa, J.F.; García-Ríos, R.I.; Cueto-Escobedo, J.; Guillen-Ruiz, G.; Bernal-Morales, B. Myristic Acid Produces Anxiolytic-Like Effects in Wistar Rats in the Elevated Plus Maze. *Biomed. Res. Int.* **2014**, *2014*. [CrossRef] [PubMed]

15. Guo, Z.; Vikbjerg, A.F.; Xu, X. Enzymatic modification of phospholipids for functional applications and human nutrition. *Biotechnol. Adv.* **2005**, *23*, 203–259. [CrossRef] [PubMed]

16. Vikbjerg, A.F.; Mu, H.; Xu, X. Lipase-catalyzed acyl exchange of soybean phosphatidylcholine in n-Hexane: A critical evaluation of both acyl incorporation and product recovery. *Biotechnol. Prog.* **2005**, *21*, 397–404. [CrossRef] [PubMed]

17. Zhao, T.; No, D.S.; Kim, B.H.; Garcia, H.S.; Kim, Y.; Kim, I.H. Immobilized phospholipase A1-catalyzed modification of phosphatidylcholine with *n*-3 polyunsaturated fatty acid. *Food Chem.* **2014**, *157*, 132–140. [CrossRef] [PubMed]

18. Chojnacka, A.; Gładkowski, W.; Grudniewska, A. Lipase-catalyzed transesterification of egg-yolk phophatidylcholine with concentrate of *n*-3 polyunsaturated fatty acids from cod liver oil. *Molecules* **2017**, *22*, 1771. [CrossRef] [PubMed]

19. Li, X.; Chen, J.F.; Yang, B.; Li, D.M.; Wang, Y.H.; Wang, W.F. Production of structured phosphatidylcholine with high content of DHA/EPA by immobilized phospholipase A1-catalyzed transesterification. *Int. J. Mol. Sci.* **2014**, *15*, 15244–15258. [CrossRef] [PubMed]

20. Xi, X.; Feng, X.; Shi, N.; Ma, X.; Lin, H.; Han, Y. Immobilized phospholipase A1-catalyzed acidolysis of phosphatidylcholine from Antarctic krill (*Euphausia superba*) for docosahexaenoic acid enrichment under supercritical conditions. *J. Mol. Catal. B Enzym.* **2016**, *126*, 46–55. [CrossRef]

21. Chojnacka, A.; Gładkowski, W.; Gliszczyńska, A.; Niezgoda, N.; Kiełbowicz, G.; Wawrzeńczyk, C. Synthesis of structured phosphatidylcholine containing punicic acid by the lipase-catalyzed transesterification with pomegranate seed oil. *Catal. Commun.* **2016**, *75*, 60–64. [CrossRef]

22. Niezgoda, N.; Gliszczyńska, A.; Gładkowski, W.; Chojnacka, A.; Kiełbowicz, G.; Wawrzeńczyk, C. Production of concentrates of CLA obtained from sunflower and safflower and their application to the lipase-catalyzed acidolysis of egg yolk phosphatidylcholine. *Eur. J. Lipid Sci. Technol.* **2016**, *118*, 1566–1578. [CrossRef]

23. Shanker Kaki, S.; Ravinder, T.; Ashwini, B.; Rao, B.V.S.K.; Prasad, R.B.N. Enzymatic modification of phosphatidylcholine with *n*-3 PUFA from silkworm oil fatty acids. *Grasas y Aceites* **2014**, *65*, e021. [CrossRef]

24. Park, C.W.; Kwon, S.J.; Han, J.J.; Rhee, J.S. Transesterification of phosphatidylcholine with eicosapentaenoic acid ethyl ester using phospholipase A 2 in organic solvent. *Biotechnol. Lett.* **2000**, *22*, 147–150. [CrossRef]

25. Doig, S.D.; Diks, R.M.M. Toolbox for exchanging constituent fatty acids in lecithins. *Eur. J. Lipid Sci. Technol.* **2003**, *105*, 359–367. [CrossRef]

26. Mustranta, A.; Forssell, P.; Aura, A.M.; Suortti, T.; Poutanen, K. Modification of phospholipids with lipases and phospholipases. *Biocatal. Biotransform.* **1994**, *9*, 181–194. [CrossRef]

27. Chojnacka, A.; Gładkowski, W.; Kiełbowicz, G.; Gliszczyńska, A.; Niezgoda, N.; Wawrzeńczyk, C. Lipase-catalyzed interesterification of egg-yolk phosphatidylcholine and plant oils. *Grasas y Aceites* **2014**, *65*, e053. [CrossRef]

28. Werbovetz, K.A.; Englund, P.T. Lipid metabolism in Trypanosoma brucei: Utilization of myristate and myristoyllysophosphatidylcholine for myristoylation of glycosyl phosphatidylinositols. *Biochem. J.* **1996**, *318*, 575–581. [CrossRef] [PubMed]

29. Wong, J.T.; Tran, K.; Pierce, G.N.; Chan, A.C.; Karmin, O.; Choy, P.C. Lysophosphatidylcholine Stimulates the Release of Arachidonic Acid in Human Endothelial Cells. *J. Biol. Chem.* **1998**, *273*, 6830–6836. [CrossRef] [PubMed]

30. Chojnacka, A.; Gładkowski, W.; Kiełbowicz, G.; Wawrzeńczyk, C. Enzymatic enrichment of egg-yolk phosphatidylcholine with α-linolenic acid. *Biotechnol. Lett.* **2009**, *31*, 705–709. [CrossRef] [PubMed]

31. Virto, C.; Adlercreutz, P. Lysophosphatidylcholine synthesis with Candida antarctica lipase B (Novozym 435). *Enzyme Microb. Technol.* **2000**, *26*, 630–635. [CrossRef]

32. Anderson, E.M.; Larsson, K.M.; Kirk, O. One biocatalyst—Many applications: The use of Candida antarctica B-lipase in organic synthesis. *Biocatal. Biotransform.* **1998**, *16*, 181–204. [CrossRef]

33. Muralidhar, R.V.; Chirumamilla, R.R.; Marchant, R.; Ramachandran, V.N.; Ward, O.P.; Nigam, P. Understanding lipase stereoselectivity. *World J. Microbiol. Biotechnol.* **2002**, *18*, 81–97. [CrossRef]

34.	Kirk, O.; Christensen, M.W. Lipases from Candida antarctica: Unique biocatalysts from a unique origin. *Org. Process Res. Dev.* **2002**, *6*, 446–451. [CrossRef]

35.	Heldt-Hansen, H.P.; Ishii, M.; Patkar, S.A.; Hansen, T.T.; Eigtved, P. A New Immobilized Positional Nonspecific Lipase for Fat Modification and Ester Synthesis. In *Biocatalysis in Agricultural Biotechnology*; Whitaker, J.R., Sonnet, P.E., Eds.; ACS Publications: Washington, DC, USA, 1989; pp. 158–172, ISBN 9780841215719.

36.	Adlercreutz, D.; Budde, H.; Wehtje, E. Synthesis of phosphatidylcholine with defined fatty acid in the *sn*-1 position by lipase-catalyzed esterification and transesterification reaction. *Biotechnol. Bioeng.* **2002**, *78*, 403–411. [CrossRef] [PubMed]

37.	Reddy, J.R.C.; Vijeeta, T.; Karuna, M.S.L.; Rao, B.V.S.K.; Prasad, R.B.N. Lipase-catalyzed preparation of palmitic and stearic acid-rich phosphatidylcholine. *J. Am. Oil Chem. Soc.* **2005**, *82*, 727–730. [CrossRef]

38.	Vikbjerg, A.F.; Peng, L.; Mu, H.; Xu, X. Continuous production of structured phospholipids in a packed bed reactor with lipase from Thermomyces lanuginosa. *J. Am. Oil Chem. Soc.* **2005**, *82*, 237–242. [CrossRef]

39.	Kim, J.H.; Yoon, S.H. Effects of organic solvents on transesterification of phospholipids using phospholipase A2 and lipase. *Food Sci. Biotechnol.* **2014**, *23*, 1207–1211. [CrossRef]

40.	Gan, L.J.; Wang, X.Y.; Yang, D.; Zhang, H.; Shin, J.A.; Hong, S.T.; Park, S.H.; Lee, K.T. Emulsifying properties of lecithin containing different fatty acids obtained by immobilized lecitase ultra-catalyzed reaction. *J. Am. Oil Chem. Soc.* **2014**, *91*, 579–590. [CrossRef]

41.	Vikbjerg, A.F.; Mu, H.; Xu, X. Parameters affecting incorporation and by-product formation during the production of structured phospholipids by lipase-catalyzed acidolysis in solvent-free system. *J. Mol. Catal. B Enzym.* **2005**, *36*, 14–21. [CrossRef]

42.	Nandi, S. Enzymatic Synthesis and Characterization of Modified Phospholipids using Decanoic Acid. *Int. J. Biotechnol. Biochem.* **2017**, *13*, 31–38.

43.	Hama, S.; Miura, K.; Yoshida, A.; Noda, H.; Fukuda, H.; Kondo, A. Transesterification of phosphatidylcholine in *sn*-1 position through direct use of lipase-producing Rhizopus oryzae cells as whole-cell biocatalyst. *Appl. Microbiol. Biotechnol.* **2011**, *90*, 1731–1738. [CrossRef] [PubMed]

44.	Egger, D.; Wehtje, E.; Adlercreutz, P. Characterization and optimization of phospholipase A2 catalyzed synthesis of phosphatidylcholine. *Biochim. Biophys. Acta Protein Struct. Mol. Enzymol.* **1997**, *1343*, 76–84. [CrossRef]

45.	Ghosh, M.; Bhattacharyya, D.K. Soy lecithin-monoester interchange reaction by microbial lipase. *J. Am. Oil Chem. Soc.* **1997**, *74*, 761–763. [CrossRef]

46.	Kiełbowicz, G.; Gładkowski, W.; Chojnacka, A.; Wawrzeńczyk, C. A simple method for positional analysis of phosphatidylcholine. *Food Chem.* **2012**, *135*, 2542–2548. [CrossRef] [PubMed]

47.	Gładkowski, W.; Chojnacka, A.; Kiełbowicz, G.; Trziszka, T.; Wawrzeńczyk, C. Isolation of Pure Phospholipid Fraction from Egg Yolk. *J. Am. Oil Chem. Soc.* **2012**, *89*, 179–182. [CrossRef]

*catalysts*

MDPI

*Article*

# Diastereoselective Synthesis of 7,8-Carvone Epoxides

Sofia Pombal [1], Ignacio E. Tobal [2], Alejandro M. Roncero [2], Jesus M. Rodilla [1],
Narciso M. Garrido [2], Francisca Sanz [3], Alberto Esteban [2], Jaime Tostado [2], Rosalina F. Moro [2],
Maria Jose Sexmero [2], Pablo G. Jambrina [4] and David Diez [2,*]

[1] Departamento de Química, Faculdade de Ciências, FibEnTech—Materiais Fibrosos e Tecnologias
Ambientais, Universidade da Beira Interior, Rua Marques de Ávila e Bolama, 6201-001 Covilhã, Portugal;
sofia.pombal@gmail.com (S.P.); rodilla@ubi.pt (J.M.R.)
[2] Departamento de Química Orgánica, Facultad de Ciencias Químicas, Universidad de Salamanca,
Plaza de los Caídos 1-5, 37008 Salamanca, Spain; ignaciotobal@usal.es (I.E.T.);
alexmaron@usal.es (A.M.R.); nmg@usal.es (N.M.G.); aesteban@usal.es (A.E.); u158198@usal.es (J.T.);
rfm@usal.es (R.F.M.); mjsex@usal.es (M.J.S.)
[3] Servicio de Difracción de Rayos X, Nucleus, Universidad de Salamanca, Plaza de los Caídos 1-5,
37008 Salamanca, Spain; sdrayosx@usal.es
[4] Departamento de Química Física Aplicada, Facultad de Ciencias, Universidad Autónoma de Madrid,
Ciudad Universitaria de Cantoblanco, 28049 Madrid, Spain; pablojambrina@gmail.com
* Correspondence: ddm@usal.es; Fax: +34-923-294574

Received: 4 June 2018; Accepted: 16 June 2018; Published: 19 June 2018

**Abstract:** The synthesis of the two 7,8-epoxides of carvone has been attained using organocatalysis
in a two-step synthetic route through a bromoester intermediate. Among the different reaction
conditions tested for the bromination reaction, moderate yields and diastereoselection are achieved
using proline, quinidine, and diphenylprolinol, yielding the corresponding bromoesters that were
transformed separately into their epoxides, obtaining the enantiopure products.

**Keywords:** organocatalysis; aminocatalysis; proline; carvone; epoxidation; epoxide

## 1. Introduction

Carvone is a natural product frequently used as starting material for the synthesis of other
naturally occurring compounds [1]. The interest in the carvone molecule as a chiral pool starting
material is based on the presence of an $\alpha,\beta$-unsaturated carbonyl in a cyclohexane ring with a
chiral centre and an extra double bond in the isopropyl chain [2–6]. Regiospecific epoxidation of
carvone has been a classical challenge in organic chemistry [7]. Epoxidation of carvone in the internal
double bond has been achieved in a diastereoselective reaction controlled by the chiral centre in basic
conditions. However, when the epoxidation of the isopropylidene group with *m*-CPBA is carried out,
no diastereoselection is obtained [8,9]. Recently, very interesting biological applications of the natural
7,8-epoxides mixture of carvone [10] have been described: of anti-inflammatory [10], antifeedant,
and phytotoxic [11] nature. To obtain these epoxy derivatives, the organic oxidants used for the
epoxidation reaction are rather expensive and have to be used in a stoichiometric manner, and even in
those conditions, the reaction yields are rather small. For these reasons, the interest in the synthesis
of the 7,8-epoxides of carvone using catalysis has increased. Many metals have been used in this
catalysed oxidation reaction: gallium [12]; aluminium [12]; vanadium and tungsten [13]; and di- or
tetracoordinated metals with diamines, mainly manganese [14–16] (Scheme 1) or iron [17–20], or
chitosan based with manganese, copper, cobalt, or nickel [21]. In all cases, the diastereoselectivity was
very small or even negligible. An alternative method used for the synthesis of the 7,8-carvone epoxides
consisted of the enzymatic oxidation of carvone by cytochrome $P450_{cin}$ [22]. In the last years, there
has been a great interest in the use of greener methods for this oxidation, and it has been achieved by

chemoenzymatic methods [23] or with hydrotalcites and hydrogen peroxide [24]. Another strategy has been afforded, using *N*-bromosuccinimide (NBS) in DMSO and ulterior addition of bases [25]. In all these cases, no diastereoselectivity was found (0% diastereomeric ratio, d.r.)

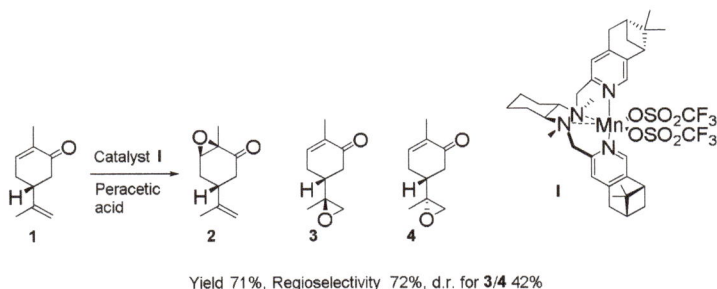

Yield 71%, Regioselectivity 72%, d.r. for **3/4** 42%

**Scheme 1.** Synthesis of the epoxides with 42 d.r.

In the last years, there has been an increasing interest in metal-free reactions, with organocatalysis being one of the most interesting areas of research in organic chemistry [26–30]. An elevated number of chiral organocatalysts has been developed, where amines play an important role and comprise the main family of organocatalysts (Figure 1).

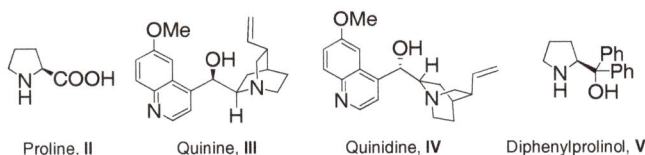

Proline, **II**      Quinine, **III**      Quinidine, **IV**      Diphenylprolinol, **V**

**Figure 1.** Organocatalysts frequently used in organic synthesis (proline, **II**; quinine, **III**; quinidine, **IV**; and diphenylprolinol, **V**).

We designed a different route based on the enantioselective halogenation of olefins by organocatalytic methods in order to obtain the required carvone epoxides. The halocyclization reactions have been a subject of interest for the organic chemist over the years [31]. Recently, the asymmetric halogenation of olefins has become an area of expansion, with many research groups interested in it [32–34].

Due to the importance of the natural epoxides **3** and **4** for the synthesis of natural products [35], where they act as starting materials, there has been a great interest in their synthesis. Notwithstanding this, to the best of our knowledge, they have not been separated or obtained separately. In this work, we report the synthesis of the 7,8-epoxyderivatives **3** and **4** separately from the bromoester derivatives of carvone **7** and **8** using an organocatalysed procedure.

## 2. Results and Discussion

Due to the different reactivity of the two double bonds of carvone, the enantioselective bromo-alcohol functionalization of the terminal double bond can be achieved using an organocatalyst in acidic conditions. It has been known for a long time that the use of amino acids with NBS produces a bromine transfer into the amino acid, which, if chiral, could be used for enantioselective bromination [36]. It is also known that quinidine and quinine derivatives form complexes with halogen that have been used in enantioselective halocyclization reactions [37]. For these reasons, we selected the organocatalysts of Figure 1 (**II–V**) to obtain the corresponding bromoderivatives, listed in Table 1. In this case, we decided to use acidic media to open the corresponding bromonium intermediate. First of all, the bromination reaction of carvone, **1**, was carried out by reaction with NBS and *o*-nitrobenzoic acid in order to facilitate the ester hydrolysis afterwards, and different organocatalysts, such as proline **II**, quinine **III**, quinidine **IV**, or diphenylprolinol **V**, using $CH_2Cl_2$ as the solvent (Table 1).

**Table 1.** Organocatalyst scope and optimization of reaction conditions for the synthesis of the bromoesters **7** and **8**.

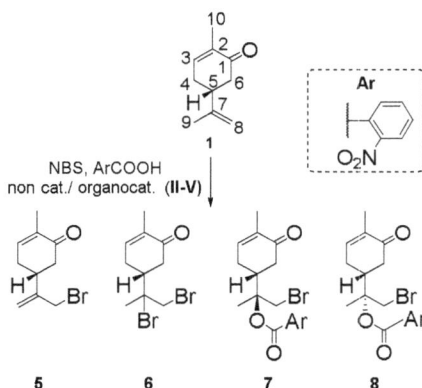

| Entry [a] | Catalyst, Load | T (°C) | Time (days) | Yield (%) [b] | | | | | d.r. (%) [c] |
|---|---|---|---|---|---|---|---|---|---|
| | | | | 1 | 5 | 6 | 7 | 8 | |
| 1 | No catalyst | rt | 6 | 15 | 28 | - | 15 | 12 | 11 |
| 2 | No catalyst | 39 | 3 | 15 | 27 | - | 18 | 16 | 6 |
| 3 | Proline, **II**, 2% | rt | 6 | - | 12 | 4 | 34 | 20 | 26 |
| 4 | Proline, **II**, 20% | rt | 6 | - | 13 | 5 | 36 | 18 | 33 |
| 5 | Proline, **II**, 20% | 39 | 6 | - | 17 | - | 34 | 18 | 31 |
| 6 | Quinine, **III**, 2% | 39 | 6 | - | 14 | - | 36 | 20 | 29 |
| 7 | Quinine, **III**, 20% | rt | 6 | - | 16 | - | 35 | 13 | 46 |
| 8 | Quinine, **III**, 20% | 39 | 6 | - | 17 | - | 36 | 16 | 38 |
| 9 | Quinidine, **IV**, 2% | rt | 6 | - | 13 | - | 56 | 11 | 67 |
| 10 | Quinidine, **IV**, 2% | 39 | 6 | - | 16 | - | 41 | 14 | 49 |
| 11 | Quinidine, **IV**, 20% | rt | 6 | - | 15 | - | 57 | 10 | 70 |
| 12 | Quinidine, **IV**, 20% | 39 | 6 | - | 11 | - | 54 | 16 | 54 |
| 13 | Diphenylprolinol, **V**, 2% | rt | 6 | - | 20 | - | 27 | 10 | 46 |
| 14 | Diphenylprolinol, **V**, 2% | 39 | 6 | - | 7 | - | 35 | 28 | 11 |
| 15 | Diphenylprolinol, **V**, 20% | rt | 6 | - | 21 | - | 50 | 17 | 50 |
| 16 | Diphenylprolinol, **V**, 20% | 39 | 6 | - | 23 | - | 44 | 24 | 29 |

[a] All reactions were carried out in $CH_2Cl_2$ at room temperature (r.t.) or heating at 39 °C. Reaction conditions ratio: *R*-carvone/NBS/*o*-nitrobenzoic acid 1/1.4/1.4. [b] Yields refer to isolated compounds; the remaining proportion to 100% comprises decomposition products. [c] Diastereoselection in d.r. was based on isolated compounds **7** and **8**.

As displayed in Table 1, when the reaction shown in Table 1 is carried out without a catalyst at room temperature (entry 1), the reaction is quite slow, giving mainly the allyl bromoderivative **5** and the mixture of the bromoesters **7** and **8** in relatively low yield, with carvone recovered in a 15% yield. To increase the reaction rate, the reaction was heated to 39 °C, finding no substantial increase in the yield of **7** or **8** (entry 2). Then, we carried out the organocatalysed reaction using proline **II**, quinine **III**, quinidine **IV**, or diphenylprolinol **V** at different temperatures and changing the load of catalyst. When proline **II** was used (entries 3–5), the yield in the required bromoesters (**7** and **8**), was increased with a moderated d.r., and no recovery of carvone was observed, also with a decrease in the quantity of the bromoderivative **5** produced. When the temperature was raised to 39 °C, the yield of the bromoesters **7** and **8** decreased, and also the diastereoselectivity. It is worth noticing that a larger amount of organocatalyst increased the diastereoselectivity moderately, but when heated, the diastereoselectivity decreased. A small quantity of a subproduct was identified as a mixture (3:1) of dibromoderivative **6** when proline was used as the organocatalyst (entries 3 and 4); due to the small amount obtained, the stereochemistry of the major diastereomer has not been determined. When quinine **III** is used as the catalyst (entries 6–8), it does not increase the yield for the obtention of **7** and **8**, but an increase in the diastereoselectivity is achieved when a 20% load is used (entry 7), leading to a global yield of 48% for the bromoesters and a diastereoselectivity of 46%. When quinidine **IV** was used, it led to an increase in the yield and diastereoselectivity (entries 9–12); in particular, if 20% of quinidine **IV** was used at room temperature (entry 11), the yield of the bromoesters rose to 67% with a diastereoselectivity of 70%. When diphenylprolinol **V** was used as the organocatalyst (entries 13–16) the best results were observed at room temperature and at a 20% load (entry 15).

To sum up, our results show that the use of an organocatalyst enhances the yield for the obtention of the bromoesters **7** and **8**, leading to moderate or good diastereocontrol, depending on the catalyst used and its load.

The structures of all the compounds were determined by their spectroscopic properties, and the stereochemistry was established by X-ray diffraction of the crystal structures of the bromoester **7** and **8** (Figure 2) [38]. In this manner, there was not any doubt about the stereochemistry of these compounds.

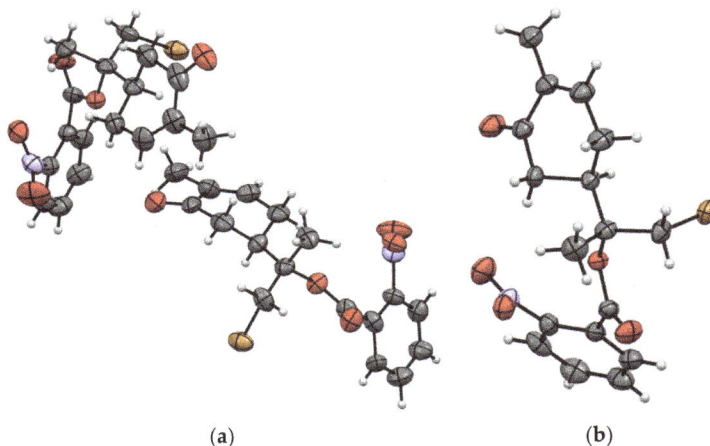

(a)          (b)

**Figure 2.** X-ray-determined structures of compounds: (a) bromoester **7** crystallised as dimers in the cell unit and (b) bromoester **8**. Atom colours: grey (C), white (H), blue (N), red (O) and light brown (Br).

To shed some light onto the mechanism of the reaction, we carried out a conformational study of the landscape associated with the dihedral angle between C8–C7–C4–C3 of *R*-carvone. In these calculations, the solvent $CH_2Cl_2$ was simulated using the polarisable continuum model (see Methods for further details). The results, depicted in Figure 3, show that there are three minima separated by low energy barriers that could be easily crossed at room temperature, as expected for the rotation along a single C–C bond. The most stable conformation corresponds to that with the side chain perpendicular to the phenyl ring and with the methyl group *anti* to the H of C5. Interestingly, the structure with the methyl group *syn* to the H of C5 corresponds to a maximum (both structures are shown in Figure 3). The structure of the three minima correspond to those found by Avilés and coworkers [39] for *S*-carvone, although there are some differences in the barrier heights that separate them, probably due to the different *ab initio* method used and the fact that no solvent was included in the simulations of the Ref. [39].

Using the *ab initio* calculations, we could estimate that at 39 °C, *R*-carvone populates 58% of the lowest energy rotamer. These results are in good agreement with the experimental results in Table 1, explaining why only a modest stereoselectivity was achieved in the absence of catalysts.

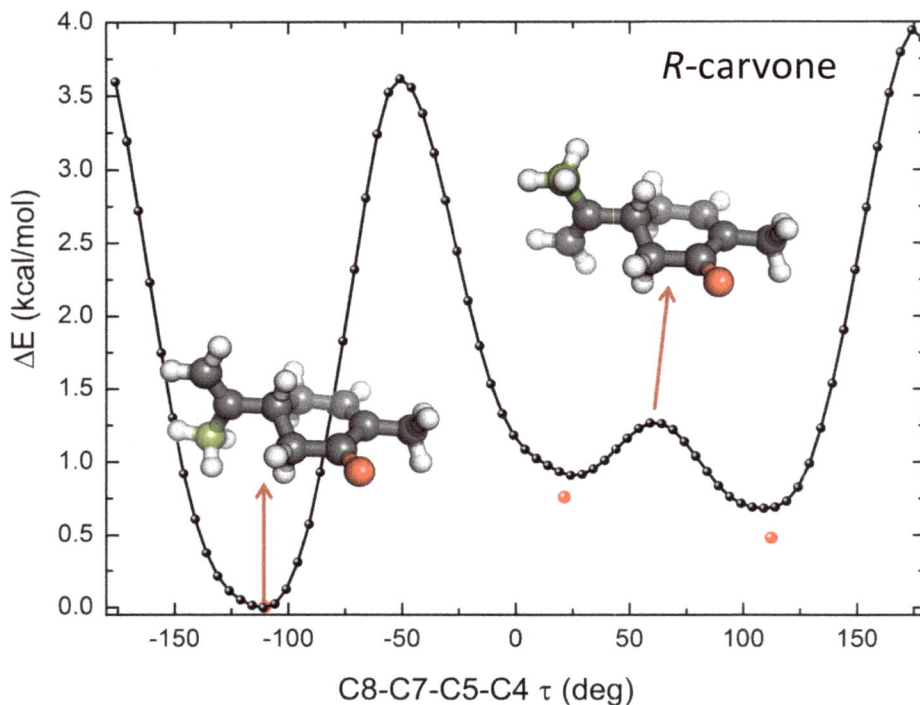

**Figure 3.** Scan of the potential energy surface calculated along the C8–C7–C4–C3 of *R*-carvone at a MP2/6-31+G(d,p) level of theory (black points). The minima were further optimized at a MP2/6-311++G(d,p) level of theory and their relative energies are also shown as red points. The structures corresponding to the global minimum and the transition state between the two local minima are shown as spheres (red (O), white (H), grey (C), to highlight the difference between both conformers C9 is shown in green). (See text for further details).

Keeping in mind this conformational analysis, we suggest the reaction mechanism displayed in Figure 4, where a hydrogen bond between the carbonyl group of carvone and the hydroxyl group of the organocatalyst stabilizes the complex **VI**, leading to a bromination of the alpha side and the entry of the nucleophile by the beta side, as shown in Figure 4.

The bromoester derivatives **7** and **8** are intermediates in the synthesis of the epoxy derivatives of carvone. The hydrolysis reaction of **7** and **8** with alkali grants the synthesis of the epoxy derivatives **3** and **4**, respectively (see Scheme 2).

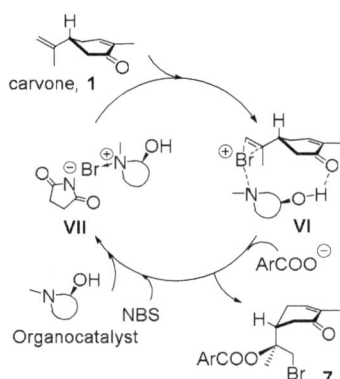

**Figure 4.** Proposed mechanism for the observed stereochemical outcome.

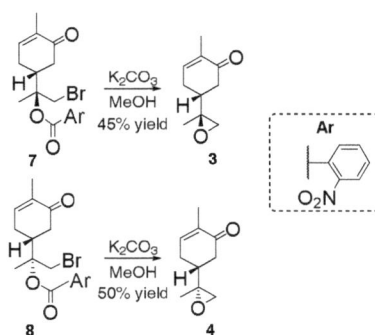

**Scheme 2.** Synthesis of 7,8-carvone epoxides **3** and **4**. Further research using *S*-carvone is in process in order to test the possible match–mismatch diastereoselectivity.

## 3. Materials and Methods

### 3.1. Reagents

Unless otherwise stated, all chemicals were purchased as the highest purity commercially available and were used without further purification. Diethyl ether, tetrahydrofuran and benzene were distilled from sodium and benzophenone under argon atmosphere; dichloromethane and pyridine were distilled from calcium hydride under argon atmosphere.

### 3.2. Characterization

Melting points are uncorrected. $^1$H and $^{13}$C NMR spectra were recorded on a Bruker Advance 400 MHz DRX (400 MHz $^1$H and 100 MHz $^{13}$C) (Bruker Biospin, Wissembourg, France) and a VARIAN 200 (200 MHz $^1$H and 50 MHz $^{13}$C) (Varian Inc, Palo Alto, California, USA). Chemical shifts are expressed in δ (ppm) and coupling constants (*J*) are given in Hz. All spectra were performed in CDCl$_3$ as solvent and referenced to the residual peak of CHCl$_3$ at δ = 7.26 ppm for $^1$H and δ = 77.0 ppm for $^{13}$C.

### X-Ray Crystallography

Single crystals of **7** and **8** for XRD (X-ray diffraction) analysis were obtained by slow evaporation of a solution in hexane/CH$_2$Cl$_2$ (98:2) at room temperature. X-ray diffraction intensity data were collected for **7** and **8** compounds on a Bruker Kappa Apex II single crystal X-ray diffractometer (Bruker

AXS Inc, Madison, Wisconsin, USA), equipped with graphite monochromator, CuK$_\alpha$ ($\lambda$ = 1.54178 Å) radiation and CCD detector. Suitable single crystals were mounted on a glass fiber using cyanoacrylate adhesive. The unit cell parameters were determined by collecting the diffracted intensities from 36 frames measured in three different crystallographic zones and using the method of difference vectors. Typical data sets consist of combinations of $\omega$ and $\phi$ scan frames with a typical scan width of 0.5° and an exposure time of 10 s/frame at a crystal-to-detector distance of ~4.0 cm. The collected frames were integrated with the SAINT software package (SAINT, Bruker AXS Inc, Madison, Wisconsin, USA) using a narrow-frame algorithm. Final cell constants were determined by the global refinement of reflections from the complete data set. Data were corrected for absorption effects using the multiscan method implemented in SADABS. Structure solutions and refinement were carried out using the SHELXTL software package (SHELXTL, Bruker AXS Inc, Madison, Wisconsin, USA). The structures were solved by direct methods combined with difference Fourier synthesis and refined by full-matrix least-squares procedures, with anisotropic thermal parameters in the last cycles of refinement for all non-hydrogen atoms. The hydrogen atoms were positioned geometrically.

### 3.3. Computational Methods

Electronic structure calculations were carried out using the Gaussian09 [40] suite of programs on *R*-carvone. The dihedral angle between C8, C7, C5, and C4 was scanned at a MP2/6-31+G(d,p) level of theory, finding three minima whose structures were further optimized at a MP2/6-311++G(d,p) level of theory. The frequencies corresponding to the three structures were calculated to ensure that they correspond to minima of the potential energy surface. Solvation effects were including using the polarizable continuum model [41] with the standard dielectric constant of 8.93 for dichloromethane. The geometry of the global minimum is given in Table 2.

To rationalize the population of *R*-carvone in the surroundings of each minima, we calculated the canonical partition function, considering that each of the aforementioned scanned points corresponded to one different state.

**Table 2.** Coordinates of the global minimum: (MP2/6-31++G(d,p), E(Eh) = −463.43093).

| | | | |
|---|---|---|---|
| C | 3.166 | −0.027 | 1.285 |
| C | 2.926 | 0.247 | −1.184 |
| C | −3.461 | −0.554 | 0.118 |
| C | −1.150 | −1.514 | −0.146 |
| C | 0.348 | −1.433 | −0.231 |
| C | 0.098 | 1.040 | −0.298 |
| C | 2.369 | 0.033 | 0.201 |
| C | −1.967 | −0.440 | −0.014 |
| C | −1.386 | 0.926 | −0.015 |
| C | 0.872 | −0.114 | 0.351 |
| O | −2.090 | 1.917 | 0.171 |
| H | 0.447 | 2.017 | 0.051 |
| H | 0.223 | 1.001 | −1.390 |
| H | −3.961 | −0.011 | −0.690 |
| H | −3.797 | −0.111 | 1.060 |
| H | −3.771 | −1.601 | 0.089 |
| H | −1.600 | −2.507 | −0.186 |
| H | 0.791 | −2.283 | 0.303 |
| H | 0.654 | −1.530 | −1.284 |
| H | 0.638 | −0.116 | 1.425 |
| H | 2.587 | 1.203 | −1.599 |
| H | 4.019 | 0.254 | −1.163 |
| H | 2.600 | −0.538 | −1.875 |
| H | 2.751 | −0.168 | 2.280 |
| H | 4.246 | 0.062 | 1.197 |

## 4. Experimental Section

*4.1. General Procedure for the Synthesis of the Bromoesters **7** and **8***

To a solution of catalyst (2% or 20%) in 20 mL of CH$_2$Cl$_2$, carvone **1** (1.00 g, 6.66 mmol), nitrobenzoic acid (1.56 g, 9.32 mmol), and *N*-bromosuccinimide (1.66 g, 9.32 mmol) were added. The reaction mixture was stirred for 6 days at room temperature or at 39 °C. After that, the solvent was evaporated and the resulting crude product was purified by chromatography on silica gel (*n*-hexane/EtOAc) to provide the isolated compounds: **5**, **6**, **7** and **8**.

Example of synthetic protocol to access diastereomers esters **7** and **8**, entry 3, Table 1:

Nitrobenzoic acid (1.56 g, 9.34 mmol) and *N*-bromosuccinimide (1.66 g, 9.38 mmol) were added to a solution containing proline **II** (15.3 mg, 0.13 mmol, 2% load) and carvone **1** (1.00 g, 6.66 mmol) in 20 ml of CH$_2$Cl$_2$. The resulting mixture was stirred for 6 days at given temperature, and then the solvent was removed under vacuum and the crude product was purified by column chromatography on silica gel (*n*-hexane/EtOAc, 7:3) to yield separately the desired bromoesters **7** (34%) and **8** (20%), d.r. 26%, and the side product was identified as the alkyl bromo derivative **5** (16%), together with the dibromoester compound **6** in a very low yield (4%).

### 4.1.1. (5*R*)-9-Bromocarvone (5)

(*n*-hexane/EtOAc, 9:1) (243.2 mg, 16%). $[\alpha]_D^{20}$ = −6.8 (*c* = 1.46 in CHCl$_3$); IR (film) ν: 3419, 2976, 2937, 1746, 1709, 1663, 1382, 1220, 1058, 950; $^1$H NMR (400 MHz CDCl$_3$,): δ = 6.72 (1H, ddd, *J* = 6.1, 3.0, 1.6 Hz, H-3), 5.26 (1H, s, Ha-8), 5.01 (1H, d, *J* = 1.3 Hz, Hb-8), 3.97 (2H, d, *J* = 2.0 Hz, H-9), 2.97 (1H, ddd, *J* = 14.4 Hz, 10.0, 4.6 Hz, H-5), 2.61 (1H, m, Ha-6), 2.53 (1H, m, Ha-4), 2.34 (1H, dd, *J* = 16.0, 12.9 Hz, Hb-6), 2.27 (1H, m, Hb-4), 1.74 (3H, s, H-10); $^{13}$C NMR (100 MHz, CDCl$_3$,): δ = 199.0, 147.0, 144.2, 135.7, 115.8, 43.3, 38.2, 35.2, 31.6, 15.8; HRMS (EI) *m*/*z* calculated for C$_{10}$H$_{14}$BrO 229.0223 (M + H$^+$) found 229.0222.

### 4.1.2. (5*R*,7*RS*)-7,8-Dibromocarvone (6)

(*n*-hexane/EtOAc, 95:5) (82.0 mg, 4%). $[\alpha]_D^{20}$ = −10.6 (*c* = 1.88 mixture of diastereomers 3/1 in CHCl$_3$); IR (film) ν: 2980, 2923, 1668, 1380, 1257, 1083, 1060, 904, 711; $^1$H NMR (400 MHz CDCl$_3$,): δ = 6.71 (1H, dt, *J* = 15.0, 4.6 Hz, H-6), 3.93 (1H, m, Ha-8), 3.82 (1H, d, *J* = 10.3 Hz, Hb-8), 2.60 (2H, m, H-6), 2.42 (3H, m, H-5), 1.84 (3H, s, H-9), 1.76 (s, 3H, H-10); $^{13}$C NMR (100 MHz, CDCl$_3$,): δ = 198.4, 143.4, 135.3, 71.0, 42.3, 40.7, 40.6, 28.8, 27.9, 15.6; HRMS (EI) *m*/*z* calculated for C$_{10}$H$_{16}$Br$_2$O 308.9484 (M + H$^+$) found 308.9479.

### 4.1.3. (5*R*,7*R*)-8-Bromo-7-(2-nitrobenzoate)carvone (7)

(*n*-hexane/EtOAc, 7:3) (898.1 mg, 34%). $[\alpha]_D^{20}$ = +41.7 (*c* = 0.48 in CHCl$_3$); IR (film) ν: 1728, 1668, 1530, 1348, 1289, 1255, 1123, 1066, 908, 733, 730; $^1$H NMR (400 MHz CDCl$_3$,): δ = 7.91 (1H, dd, *J* = 8.0, 1.2 Hz, H-6′), 7.72 (1H, m, H-5′), 7.64 (1H, m, H-4′), 7.61 (1H, m, H-3′), 6.74 (1H, m, H-3), 4.06 (1H, d, *J* = 11.1 Hz, Ha-8), 4.01 (1H, d, *J* = 11.1 Hz, Hb-8), 2.82 (1H, m, H-5), 2.50 (1H, m, Ha-6), 2.43 (1H, m, Ha-4), 2.29 (1H, m, Hb-6), 2.20 (1H, dd, *J* = 15.9, 14.2 Hz, Hb-4), 1.74 (s, 3H, H-10), 1.72 (s, 3H, H-9); $^{13}$C NMR (100 MHz, CDCl$_3$,): δ = 198.1, 164.4, 147.3, 144.5, 135.3, 133.4, 131.6, 129.7, 128.5, 124.0, 85.4, 41.0, 38.5, 35.4, 26.1, 19.0, 15.6; HRMS (EI) *m*/*z* calculated for C$_{17}$H$_{18}$BrNO$_5$Na 418.0260 (M + Na$^+$) found 418.0265.

Crystal data for **7**: C$_{17}$H$_{18}$BrNO$_5$, M = 396.23, monoclinic, space group P$_{21}$ (n° 4), a = 8.0408(6) Å, b = 28.336(2) Å, c = 8.1290(6) Å, α = γ = 90°, β = 103.603(4)°, V = 1800.2(2) Å3, Z = 4, D$_c$ = 1.462 Mg/m$^3$, m = (Cu-K$_\alpha$) = 3.340 mm$^{-1}$, F(000) = 808. 10420 reflections were collected at 3.12 ≤ 2θ ≤ 67.40 and merged to give 4916 unique reflections (R$_{int}$ = 0.0872), of which 2992 with I > σ (I) were considered to be observed. Final values are R = 0.0765, *w*R = 0.2113, GOF = 1.026, max/min residual electron density 0.703 and −0.652 e. Å$^{-3}$.

### 4.1.4. (5*R*,7*S*)-8-Bromo-7-(2-nitrobenzoate)carvone (8)

(*n*-hexane/EtOAc, 7:3) (528.3 mg, 20%). $[\alpha]_D^{20}$ = +6.9 (*c* = 1.44 in CHCl₃); IR (film) ν: 1728, 1668, 1530, 1348, 1289, 1255, 1123, 1066, 908, 733, 711; ¹H NMR (400 MHz CDCl₃,): δ = 7.91 (1H, d, *J* = 7.7 Hz, H-6′), 7.66 (1H, m, H-5′), 7.65 (1H, m, H-4′), 7.61 (1H, m, H-3′), 6.68 (1H, m, H-3), 4.15 (1H, d, *J* = 11.2 Hz, Ha-8), 3.90 (1H, d, *J* = 11.2 Hz, Hb-8), 2.82 (1H, m, H-5), 2.58 (1H, m, Ha-6), 2.36 (1H, m, Ha-4), 2.24 (1H, dd, *J* = 15.8, 14.7 Hz, Hb-6), 2.19 (1H, m, Hb-4), 1.72 (3H, s, H-9), 1.71 (3H, s, H-10); ¹³C NMR (100 MHz, CDCl₃,): δ = 198.5, 164.4, 147.1, 143.4, 135.6, 133.4, 133.1, 129.5, 128.5, 124.0, 85.2, 40.9, 38.2, 35.5, 26.3, 18.7, 15.6; HRMS (EI) *m/z* calculated for $C_{17}H_{18}BrNO_5Na$ 418.0266 (M + Na⁺) found 418.0265. See Figure 1 for X-ray.

Crystal data for **8**: $C_{17}H_{18}BrNO_5$, M = 396.22, monoclinic, space group P2₁ (n° 4), a = 7.7282(4) Å, b = 13.9571(11) Å, c = 8.5804(6) Å, α = γ = 90°, β = 109.037(5)°, V = 874.89(11) Å³, Z = 2, D$_c$ = 1.504 Mg/m³, m = (Cu-K$_\alpha$) = 3.437 mm⁻¹, F(000) = 404. 3704 reflections were collected at 5.45 ≤ 2θ ≤ 65.91 and merged to give 2154 unique reflections (R$_{int}$ = 0.0351), of which 1799 with I > σ (I) were considered to be observed. Final values are R = 0.0491, *w*R = 0.1280, GOF = 1.062, max/min residual electron density 0.417 and −0.716 e. Å⁻³.

### 4.2. Hydrolysis Reaction of Bromoesters 7 and 8

In a round-bottom flask equipped with a reflux condenser, compound **7** or **8**, MeOH (20 mL), and K₂CO₃ (0.1 mmol) were added. The resulting mixture was stirred at 35 °C and the progress of the reaction was monitored by thin-layer chromatography (TLC). The solvent was evaporated, and finally, the reaction product was purified by column chromatography on silica gel (*n*-hexane/EtOAc, 8:2) to obtain compound **3** (45%) or **4** (50%), respectively.

#### 4.2.1. (5*R*,7*S*)-7,8-Epoxycarvone (3)

$[\alpha]_D^{20}$ = −13.5 (*c* = 0.74 in CHCl₃); IR (film) ν: 2980, 2924, 1667, 1450, 1382, 1106, 901, 834, 704; ¹H NMR (400 MHz CDCl₃,): δ = 6.72 (m, 1H, H-3), 2.65 (d, *J* = 4.6 Hz, 1H, Ha-8), 2.61 (d, *J* = 4.6 Hz, 1H, Hb-8), 2.58 (m, 1H, Ha-6), 2.51 (m, 1H, Hb-6), 2.38 (m, 1H Ha-4), 2.25 (m, 1H, Hb-4), 2.06 (m, 1H, H-5), 1.77 (s, 3H, H-10), 1.29 (s, 3H, H-9); ¹³C NMR (100 MHz, CDCl₃,): δ = 198.2, 144.2, 135.6, 58.0, 52.9, 41.3, 40.4, 27.7, 18.4, 15.7; HRMS (EI) *m/z* calculated for $C_{10}H_{14}O_2Na$ 189.0886 (M + Na⁺) found 189.0891.

#### 4.2.2. (5*R*,7*R*)-7,8-Epoxycarvone (4)

$[\alpha]_D^{20}$ = +12.8 (*c* = 0.74 in CHCl₃); IR (film) ν: 2980, 2924, 1667, 1450, 1382, 1106, 901, 834, 704; ¹H NMR (400 MHz CDCl₃,): δ = 6.73 (m, 1H, H-3), 2.72 (d, *J* = 4.5 Hz, 1H, Ha-8), 2.58 (d, *J* = 4.5 Hz, 1H, Hb-8), 2.55 (m, 1H, Ha-6), 2.06 (m, 1H, H-5), 2.17 (m, 3H, Hb-6, Ha-4, Hb-4) 1.76 (s, 3H, H-10), 1.31 (s, 3H, H-9); ¹³C NMR (100 MHz, CDCl₃,): δ = 199.1, 144.1, 135.6, 57.9, 52.4, 40.7, 39.9, 27.9, 19.0, 15.7; HRMS (EI) *m/z* calculated for $C_{10}H_{14}O_2Na$ 189.0886 (M + Na⁺) found 189.0891.

## 5. Conclusions

Throughout this article, we have developed a novel method for the selective synthesis of 7,8-carvone epoxy derivatives. The diastereoselective synthesis of these epoxides from carvone in two steps with bromoester intermediates using organocatalysis has been achieved. To the best of our knowledge, it is the first time that these epoxides have been obtained separately; and due to their synthetic potential, this methodology can be used for the enantioselective synthesis of many natural products.

**Author Contributions:** Conceptualization: A.M.R and D.D.; investigation: S.P., I.E.T., A.M.R. and J.M.R.; methodology: S.P., N.M.G., R.F.M., M.J.S.; supervision: D.D. and J.M.R.; writing of original draft: A.M.R., I.E.T., A.E., J.T.; molecular modelling: P.G.J.; X-ray: F.S.

**Funding:** Financial support for this work came from the Ministry of Economy and Competitiveness (MINECO/FEDER-CTQ2015-68175-R), the European Regional Development Fund (FEDER), the Regional

Government of Castilla y León (BIO/SA59/15, UIC21) and the Universidad de Salamanca. AMR and IET thank the Ministry of Education, Culture and Sports (MECD) and the Regional Government of Castile & Leon for their fellowships, respectively. P.G.J. gratefully acknowledges funding by the Spanish Ministry of Science and Innovation (grant MINECO/FEDER-CTQ2015-65033-P) and computing time allocation from the "Centro de Computación Científica" at Universidad Autónoma de Madrid.

**Acknowledgments:** The authors thank also Anna M. Lithgow for the NMR spectra and César Raposo for the mass spectra. This manuscript is dedicated to Isidro S. Marcos on the occasion of his 65th birthday.

**Conflicts of Interest:** The authors declare no conflict of interest. The funders had no role in the design of the study; in the collection, analyses, or interpretation of data; in the writing of the manuscript; or in the decision to publish the results.

## References and Note

1. Finkbeiner, P.; Murai, K.; Röpke, M.; Sarpong, R. Total synthesis of terpenoids employing a "benzannulation of carvone" strategy: Synthesis of (−)-crotogoudin. *J. Am. Chem. Soc.* **2017**, *139*, 11349–11352. [CrossRef] [PubMed]

2. Nannini, L.J.; Nemat, S.J.; Carreira, E.M. Total synthesis of (+)-sarcophytin. *Angew. Chem. Int. Ed.* **2018**, *57*, 823–826. [CrossRef] [PubMed]

3. Kobayashi, K.; Kunimura, R.; Takagi, H.; Hirai, M.; Kogen, H.; Hirota, H.; Kuroda, C. Total synthesis of highly oxygenated bisabolane sesquiterpene isolated from ligularia lankongensis: Relative and absolute configurations of the natural product. *J Org Chem* **2018**, *83*, 703–715. [CrossRef] [PubMed]

4. Senthil Kumaran, R.; Mehta, G. A versatile, rcm based approach to eudesmane and dihydroagarofuran sesquiterpenoids from (−)-carvone: A formal synthesis of (−)-isocelorbicol. *Tetrahedron* **2015**, *71*, 1718–1731. [CrossRef]

5. Abad, A.; Agulló, C.; Cuñat, A.C.; Llosá, M.C. Stereoselective construction of the tetracyclic scalarane skeleton from carvone. *Chem. Commun.* **1999**, 427–428. [CrossRef]

6. Abad, A.; Agulló, C.; Cuñat, A.; De Alfonso, I.; Navarro, I.; Vera, N. Synthesis of highly functionalised enantiopure bicyclo[3.2.1]-octane systems from carvone. *Molecules* **2004**, *9*, 287–299. [CrossRef] [PubMed]

7. Mak, K.K.W.; Lai, Y.M.; Siu, Y.-H. Regiospecific epoxidation of carvone: A discovery-oriented experiment for understanding the selectivity and mechanism of epoxidation reactions. *J. Chem. Ed.* **2006**, *83*, 1058. [CrossRef]

8. Masarwa, A.; Weber, M.; Sarpong, R. Selective c–c and c–h bond activation/cleavage of pinene derivatives: Synthesis of enantiopure cyclohexenone scaffolds and mechanistic insights. *J. Am. Chem. Soc.* **2015**, *137*, 6327–6334. [CrossRef] [PubMed]

9. Valeev, R.F.; Bikzhanov, R.F.; Selezneva, N.K.; Gimalova, F.A.; Miftakhov, M.S. Synthesis of 6-hydroxycarvone derivatives and their oxidative decyclization with lead tetraacetate. *Russ. J. Org. Chem.* **2011**, *47*, 1287. [CrossRef]

10. Sepúlveda-Arias, J.C.; Veloza, L.A.; Escobar, L.M.; Orozco, L.M.; Lopera, I.A. Anti-inflammatory effects of the main constituents and epoxides derived from the essential oils obtained from tagetes lucida, cymbopogon citratus, lippia alba and eucalyptus citriodora. *J. Essent. Oil Res.* **2013**, *25*, 186–193. [CrossRef]

11. Kimbaris, A.C.; González-Coloma, A.; Andrés, M.F.; Vidali, V.P.; Polissiou, M.G.; Santana-Méridas, O. Biocidal compounds from mentha sp. Essential oils and their structure–activity relationships. *Chem. Biodivers.* **2017**, *14*. [CrossRef] [PubMed]

12. Mandelli, D.; Kozlov, Y.N.; da Silva, C.A.R.; Carvalho, W.A.; Pescarmona, P.P.; Cella, D.d.A.; de Paiva, P.T.; Shul'pin, G.B. Oxidation of olefins with $H_2O_2$ catalysed by gallium(III) nitrate and aluminum(III) nitrate in solution. *J. Mol. Catal. A Chem.* **2016**, *422*, 216–220. [CrossRef]

13. Kamata, K.; Sugahara, K.; Yonehara, K.; Ishimoto, R.; Mizuno, N. Efficient epoxidation of electron-deficient alkenes with hydrogen peroxide catalysed by $[\gamma\text{-}PW_{10}O_{38}V_2(\mu\text{-}OH)_2]^{3-}$. *Chemistry* **2011**, *17*, 7549–7559. [CrossRef] [PubMed]

14. Gomez, L.; Garcia-Bosch, I.; Company, A.; Sala, X.; Fontrodona, X.; Ribas, X.; Costas, M. Chiral manganese complexes with pinene appended tetradentate ligands as stereoselective epoxidation catalysts. *Dalton Trans.* **2007**, 5539–5545. [CrossRef] [PubMed]

15. Yu, S.; Miao, C.-X.; Wang, D.; Wang, S.; Xia, C.; Sun, W. Mnii complexes with tetradentate n4 ligands: Highly efficient catalysts for the epoxidation of olefins with $H_2O_2$. *J. Mol. Catal. A Chem.* **2012**, *353–354*, 185–191. [CrossRef]

16. Garcia-Bosch, I.; Ribas, X.; Costas, M. A broad substrate-scope method for fast, efficient and selective hydrogen peroxide-epoxidation. *Adv. Syn. Catal.* **2009**, *351*, 348–352. [CrossRef]

17. Spannring, P.; Yazerski, V.A.; Chen, J.; Otte, M.; Weckhuysen, B.M.; Bruijnincx, P.C.A.; Klein Gebbink, R.J.M. Regioselective cleavage of electron-rich double bonds in dienes to carbonyl compounds with [Fe(OTf)$_2$(mix-BPBP)] and a combination of $H_2O_2$ and NaIO$_4$. *Eur. J. Inorg. Chem.* **2015**, *2015*, 3462–3466. [CrossRef]

18. Oddon, F.; Girgenti, E.; Lebrun, C.; Marchi-Delapierre, C.; Pécaut, J.; Ménage, S. Iron coordination chemistry of $N_2Py_2$ ligands substituted by carboxylic moieties and their impact on alkene oxidation catalysis. *Eur. J. Inorg. Chem.* **2012**, *2012*, 85–96. [CrossRef]

19. Clemente-Tejeda, D.; López-Moreno, A.; Bermejo, F.A. Non-heme iron catalysis in C=C, C–H, and $CH_2$ oxidation reactions. Oxidative transformations on terpenoids catalysed by Fe(bpmen)(OTf)$_2$. *Tetrahedron* **2013**, *69*, 2977–2986. [CrossRef]

20. Yazerski, V.A.; Spannring, P.; Gatineau, D.; Woerde, C.H.M.; Wieclawska, S.M.; Lutz, M.; Kleijn, H.; Klein Gebbink, R.J.M. Making Fe(BPBP)-catalysed C-H and C=C oxidations more affordable. *Org. Biomol. Chem.* **2014**, *12*, 2062–2070. [CrossRef] [PubMed]

21. Thatte, C.S.; Rathnam, M.V.; Kumar, M.S.S. Synthesis, characterization and application of chitosan based schiff base-transition metal complexes (Mn, Cu, Co, Ni). *JOAC* **2013**, *2*, 1192.

22. Stok, J.E.; Yamada, S.; Farlow, A.J.; Slessor, K.E.; De Voss, J.J. Cytochrome P450$_{cin}$ (CYP176A1) D241N: Investigating the role of the conserved acid in the active site of cytochrome P450s. *Biochim. Biophys. Acta* **2013**, *1834*, 688–696. [CrossRef] [PubMed]

23. Méndez-Sánchez, D.; Ríos-Lombardía, N.; Gotor, V.; Gotor-Fernández, V. Chemoenzymatic epoxidation of alkenes based on peracid formation by a rhizomucor miehei lipase-catalysed perhydrolysis reaction. *Tetrahedron* **2014**, *70*, 1144–1148. [CrossRef]

24. Rodilla, J.M.; Neves, P.P.; Pombal, S.; Rives, V.; Trujillano, R.; Díez, D. Hydrotalcite catalysis for the synthesis of new chiral building blocks. *Nat. Prod. Res.* **2016**, *30*, 834–840. [CrossRef] [PubMed]

25. Majetich, G.; Shimkus, J.; Li, Y. Epoxidation of olefins by β-bromoalkoxydimethylsulfonium ylides. *Tetrahedron Lett.* **2010**, *51*, 6830–6834. [CrossRef]

26. Dalko, P.I.; Moisan, L. Enantioselective organocatalysis. *Angew. Chem. Int. Ed.* **2001**, *40*, 3726–3748. [CrossRef]

27. List, B. Asymmetric aminocatalysis. *Synlett* **2001**, *2001*, 1675–1686. [CrossRef]

28. Brown, S.P.; Brochu, M.P.; Sinz, C.J.; MacMillan, D.W.C. The direct and enantioselective organocatalytic α-oxidation of aldehydes. *J. Am. Chem. Soc.* **2003**, *125*, 10808–10809. [CrossRef] [PubMed]

29. Pidathala, C.; Hoang, L.; Vignola, N.; List, B. Direct catalytic asymmetric enolexo aldolizations. *Angew. Chem. Int. Ed.* **2003**, *42*, 2785–2788. [CrossRef] [PubMed]

30. MacMillan, D. *Asymmetric Organocatalysis: From Biomimetic Concepts to Applications in Asymmetric Synthesis*; Wiley-VCH: Weinheim, Germany, 2005; p. 454.

31. Jiang, X.; Liu, H. 4.07 electrophilic cyclization a2-knochel, paul. In *Comprehensive Organic Synthesis II*, 2nd ed.; Elsevier: Amsterdam, The Netherlands, 2014; pp. 412–494.

32. Bar, S. Organocatalysis in the stereoselective bromohydrin reaction of alkenes. *Can. J. Chem.* **2010**, *88*, 605–612. [CrossRef]

33. Castellanos, A.; Fletcher, S.P. Current methods for asymmetric halogenation of olefins. *Chem. Eur. J.* **2011**, *17*, 5766–5776. [CrossRef] [PubMed]

34. Tan, C.K.; Yeung, Y.-Y. Recent advances in stereoselective bromofunctionalization of alkenes using n-bromoamide reagents. *Chem. Commun.* **2013**, *49*, 7985–7996. [CrossRef] [PubMed]

35. Valeev, R.F.; Bikzhanov, R.F.; Yagafarov, N.Z.; Miftakhov, M.S. Synthesis of the northern fragment of an epothilone d analogue from (−)-carvone. *Tetrahedron* **2012**, *68*, 6868–6872. [CrossRef]

36. Antelo, J.M.; Arce, F.; Crugeiras, J. Kinetics of electrophilic bromine transfer from n-bromosuccinimide to amines and amino acids. *J. Chem. Soc. Perk. Trans.* **1995**, *2*, 2275–2279. [CrossRef]

37. Chen, G.; Ma, S. Enantioselective halocyclization reactions for the synthesis of chiral cyclic compounds. *Angew. Chem. Int. Ed.* **2010**, *49*, 8306–8308. [CrossRef] [PubMed]

38. Crystallographic data for the structures reported in this paper has been deposited at the Cambridge Crystallographic Data Centre as supplementary material No. CCDC 1821654–1821653 the data can be obtained free of charge from the Cambridge Crystallographic Data Centre via www.ccdc.cam.ac.uk/getstructures.

39. Avilés Moreno, J.R.; Huet, T.R.; González, J.J.L. Conformational relaxation of s-(+)-carvone and r-(+)-limonene studied by microwave fourier transform spectroscopy and quantum chemical calculations. *Struct. Chem.* **2013**, *24*, 1163–1170. [CrossRef]

40. Frisch, M.J.; Trucks, G.W.; Schlegel, H.B.; Scuseria, G.E.; Robb, M.A.; Cheeseman, J.R.; Scalmani, G.; Barone, V.; Petersson, G.A.; Nakatsuji, H.; et al. *Gaussian 09 rev. B.01*; Gaussian Inc.: Wallingford, CT, USA, 2009.

41. Scalmani, G.; Frisch, M.J. Continuous surface charge polarizable continuum models of solvation. I. General formalism. *J. Chem. Phys.* **2010**, *132*, 114110. [CrossRef] [PubMed]

*catalysts*

MDPI

*Article*

# Deep Eutectic Mixtures as Reaction Media for the Enantioselective Organocatalyzed α-Amination of 1,3-Dicarbonyl Compounds

**Diego Ros Ñíguez, Pegah Khazaeli, Diego A. Alonso \* and Gabriela Guillena \***

Departamento de Química Orgánica and Instituto de Síntesis Orgánica (ISO), Facultad de Ciencias, Universidad de Alicante, Apdo. 99, E-03080 Alicante, Spain; diego.ros@ua.es (D.R.Ñ.); pkhazaeli1@hotmail.com (P.K.)
\* Correspondence: diego.alonso@ua.es (D.A.A.); gabriela.guillena@ua.es (G.G.); Tel.: +34-965909841 (D.A.A.); +34-965902888 (G.G.)

Received: 27 April 2018; Accepted: 16 May 2018; Published: 18 May 2018

**Abstract:** The enantioselective α-amination of 1,3-dicarbonyl compounds has been performed using deep eutectic solvents (DES) as a reaction media and chiral 2-amino benzimidazole-derived compounds as a catalytic system. With this procedure, the use of toxic volatile organic compounds (VOCs) as reaction media is avoided. Furthermore, highly functionalized chiral molecules, which are important intermediates for the natural product synthesis, are synthetized by an efficient and stereoselective protocol. Moreover, the reaction can be done on a preparative scale, with the recycling of the catalytic system being possible for at least five consecutive reaction runs. This procedure represents a cheap, simple, clean, and scalable method that meets most of the principles to be considered a green and sustainable process.

**Keywords:** asymmetric organocatalysis; α-amination; benzimidazole; deep eutectic solvents; natural products; green chemistry

## 1. Introduction

Asymmetric organocatalysis is an extremely attractive methodology for the preparation of functionalized chiral molecules and natural products, since small organic compounds are used as catalysts under very mild and simple reaction conditions [1–3]. Due to the lack of a metal element in the catalyst, organocatalytic methods are often used to prepare compounds that do not tolerate metal contamination such as pharmaceutical products. Asymmetric organocatalysis has become such an effective method of maintaining sustainability in organic synthesis as it provides many advantages, such as accessibility, low molecular weight, inexpensive catalysts and reduced toxicity.

Among the limited number of available green solvents, [4,5] deep eutectic solvents (DES) [6–12] maintain consistency within different criteria, such as availability, non-toxicity, inexpensiveness, high recyclability and low volatility. A deep eutectic solvent is a mixture between two or more components, one acting as hydrogen bond acceptor and the other as donor, having a melting point lower than the melting point of each one of the components. This behavior is due to hydrogen bond interactions between the acceptor and donor species. The use of DES as a reaction media is considered a new and expanding topic, which further assists and advances the importance of green chemistry. Recently, the association of these reaction media with asymmetric organocatalyzed processes [13,14] has been envisaged as a new and bright approach to advance sustainable processes.

Additionally, significant developments have been reached in the asymmetric electrophilic α-amination of carbonyl compounds through metal- or organo-catalyzed processes during recent years [15–19]. In fact, chiral carbonyl derivatives bearing stereogenic α-amine substitution are widely distributed among pharmaceutically active compounds. In particular, the organocatalyzed asymmetric α-amination of prochiral 1,3-dicarbonyl compounds have received great interest, since the resulting functionalized chiral molecules can be further elaborated allowing the synthesis of chiral biologically active natural products. However, this process remains unexplored with DES as reaction media.

Our research group has established the practicality of bifunctional chiral 2-aminobenzimidazole derivatives [20,21] **1** and **3** (Scheme 1) as efficient organocatalysts in the asymmetric conjugate addition of 1,3-dicarbonyl compounds to nitroolefins [22] and maleimides [23,24] as well as in the α-functionalization [25–28] of these interesting nucleophiles using volatile organic solvents (VOCs) as a reaction medium. More fascinating, we have also demonstrated that the catalytic system based on the deep eutectic solvent choline chloride/glycerol and chiral 2-aminobenzimidazole organocatalysts **2** efficiently promotes the enantioselective addition of 1,3-dicarbonyl compounds to β-nitrostyrenes, avoiding the use of toxic VOC as reaction media [29].

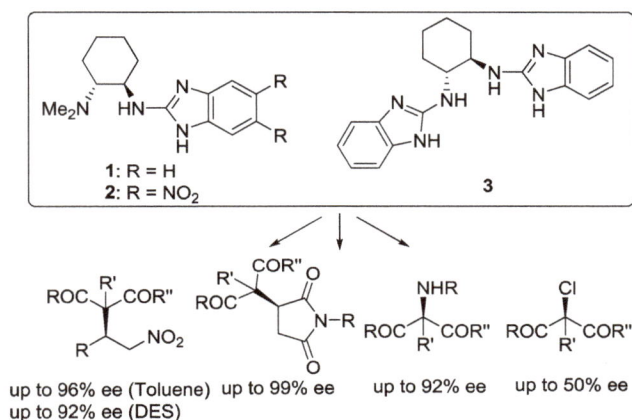

**Scheme 1.** Chiral benzimidazoles in asymmetric organocatalysis.

For asymmetric organocatalyzed processes, the use of DES as a reaction medium has been barely studied, with the aldol reaction [30–34] and conjugated addition [13,29,35] being the main focus. For these processes a rational design of the organocatalyst and the right choice of the DES has shown to be critical to obtain good results and allow organocatalyst recycling.

Herein, the use of chiral benzimidazole derivatives as organocatalysts for the electrophilic α-amination of 1,3-dicarbonyl compounds using DES as reaction media is presented.

## 2. Results and Discussion

Initially, the electrophilic α-amination of ethyl 2-oxocyclopentane-1-carboxylate with di-*tert*-butyl azodicarboxylate (DBAB) in the presence of catalyst 1 (10 mol %) in different choline chloride-based DES was investigated at 25 and 0 °C (Table 1). In general, good conversions and higher enantioselectivities were obtained at 0 °C, especially when using ChCl/urea (94%, 78% ee) and ChCl/glycerol (94%, 80% ee) as reaction media (Table 1, entries 2 and 6). The reaction time of the reaction was initially 4–5 h, however it was reduced to 1 h using ultrasounds (360 W) at 25 °C. The reduction in reaction time by the use of ultrasounds was previously observed in other related systems (i.e., ionic liquids), being attributed to physical-chemical effects [36,37]. As shown in Table 1, entries 12 and 13, under these conditions compound **4** was obtained with similar enantioselectivities with only a small erosion of the reaction conversion.

**Table 1.** Asymmetric α-amination of ethyl 2-oxocyclopentane-1-carboxylate with di-*tert*-butyl azodicarboxylate (DBAB). Deep eutectic solvents (DES) study.

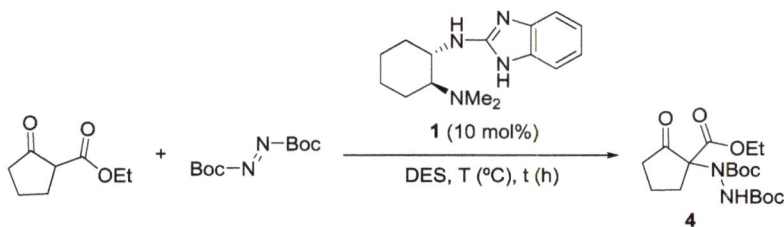

| Entry | DES | T (°C) | t (h) | Conversion (%) [1] | Ee (%) [2] |
|-------|-----|--------|-------|--------------------|-----------|
| 1 | ChCl/Urea: 1/2 | 25 | 5 | 61 | 77 |
| 2 | ChCl/Urea: 1/2 | 0 | 5 | 94 | 78 |
| 3 | AcChCl/Urea: 1/2 | 25 | 5 | 64 | 72 |
| 4 | AcChCl/Urea: 1/2 | 0 | 5 | 55 | 75 |
| 5 | ChCl/Glycerol: 1/2 | 25 | 5 | 94 | 73 |
| 6 | ChCl/Glycerol: 1/2 | 0 | 5 | 94 | 80 |
| 7 | ChCl/Ethyleneglycol: 1/2 | 25 | 5 | 78 | 72 |
| 8 | ChCl/Ethyleneglycol: 1/2 | 0 | 5 | 84 | 70 |
| 9 | ChCl/Malic acid: 1/1 | 25 | 5 | <5 | nd |
| 10 | ChCl/Tartaric acid: 1/1 | 25 | 5 | 64 | 76 |
| 11 | ChCl/Tartaric acid: 1/1 | 0 | 5 | 58 | 77 |
| 12 | ChCl/Urea: 1/2 | 25 [3] | 1 | 92 | 76 |
| 13 | ChCl/Glycerol: 1/2 | 25 [3] | 1 | 80 | 80 |

[1] Reaction conversion towards **4** determined by GC analysis. [2] Enantiomeric excess determined by chiral HPLC analysis. [3] Reaction performed under ultrasounds irradiation (360 W).

The optimization of the reaction medium resulted in the understanding that choline chloride/glycerol and choline chloride/urea were the best solvents to go forward with the conditions study using ultrasound irradiation at 25 °C. Next, the influence of the catalyst structure in the reaction results was studied: for this purpose, a series of several chiral benzimidazole-derived organocatalysts (as well other type of organocatalysts such as thiourea or sulphonamide derivatives, see Figure S1 in Supplementary Material) were tested in the α-amination model reaction under the optimized conditions using choline chloride/urea and choline chloride/glycerol as reaction media.

In both solvents, all chiral catalysts tested showed high performance achieving high reaction conversions (70–95%). However, different results concerning the enantioselectivities were encountered depending on the steric and/or electronic nature of the chiral organocatalyst. Chiral derivative **2**, containing two strong electron-withdrawing nitro groups lead to the best results, giving product **4** in 84% ee in ChCl/urea and 82% ee in ChCl/glycerol (Table 2, entries 3 and 4). The presence of two nitro groups on the benzimidazole ring increases the hydrogen-bonding ability of **2** and as a consequence the interaction with the DES structure, leading to an improvement of the selectivity of the electrophilic amination. Conversely, a strong decrease in the enantioselectivity of the process was observed when using the sterically congested $C_2$-symmetric chiral benzimidazoles **3** and **6**, which afforded compound **4** with enantiomeric excess ranging from 33 to 44% (Table 2, entries 7–10). It can be concluded that in order to obtain good selectivity in the amination addition, it is crucial to have good correlation between the steric and electronic properties within the organocatalyst [38].

**Table 2.** Asymmetric α-amination of ethyl 2-oxocyclopentane-1-carboxylate with DBAB. Catalyst study.

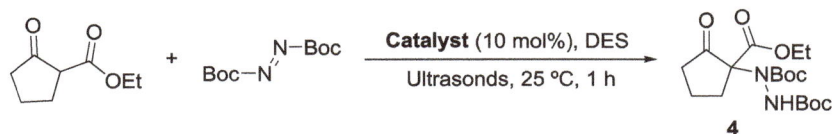

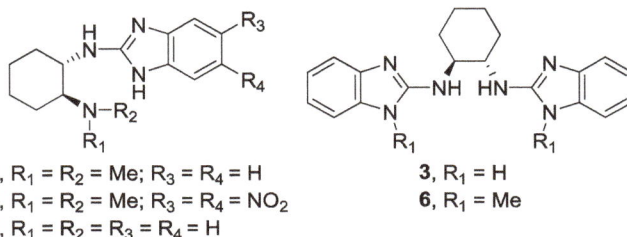

**1**, $R_1 = R_2 = Me$; $R_3 = R_4 = H$
**2**, $R_1 = R_2 = Me$; $R_3 = R_4 = NO_2$
**5**, $R_1 = R_2 = R_3 = R_4 = H$

**3**, $R_1 = H$
**6**, $R_1 = Me$

| Entry | Catalyst | DES | Conversion (%) [1] | Ee (%) [2] |
|-------|----------|-----|--------------------|------------|
| 1 | 1 | ChCl/Urea: 1/2 | 92 | 76 |
| 2 | 1 | ChCl/Glycerol: 1/2 | 80 | 80 |
| 3 | 2 | ChCl/Urea: 1/2 | 85 | 84 |
| 4 | 2 | ChCl/Glycerol: 1/2 | 90 | 82 |
| 5 | 5 | ChCl/Urea: 1/2 | 95 | 74 |
| 6 | 5 | ChCl/Glycerol: 1/2 | 70 | 78 |
| 7 | 3 | ChCl/Urea: 1/2 | 90 | 40 |
| 8 | 3 | ChCl/Glycerol: 1/2 | 95 | 44 |
| 9 | 6 | ChCl/Urea: 1/2 | 92 | 40 |
| 10 | 6 | ChCl/Glycerol: 1/2 | 91 | 33 |

[1] Reaction conversion towards **4** determined by GC analysis. [2] Enantiomeric excess determined by chiral HPLC analysis.

The recyclability of organocatalyst **2** and the eutectic liquid was performed in the model reaction under the optimized reaction conditions (Scheme 2). Therefore, to separate the DES/chiral organocatalyst mixture from the unreactive reagents and reaction products, hexane and cyclopentyl methyl ether were tested as extractive media. As shown in Scheme 2, the product was extracted and most of the catalyst remained in the DES, with the mixture being recovered and reused in five consecutive reaction runs, maintaining high enantioselectivity but with a decreased activity. Furthermore, vigorous stirring is mandatory when performing the extraction of the products to obtain a good recyclability results. For instance, in the second cycle of the cyclopentyl methyl ether recovering sequence (Scheme 2), a standard stirring was used and therefore a decrease in the conversion was observed. However, for the third run, again a vigorous stirring was applied and the conversion of the process was almost recovered.

The efficiency and synthetic utility of **2** in ChCl/glycerol was further evaluated by performing a gram-scale experiment (4.3 mmol of ethyl 2-oxocyclopentane-1-carboxylate) for the synthesis of compound **4** which was obtained in a 95% yield and 85% ee (Scheme 3).

Lastly, the influence of different electrophiles and nucleophiles were assessed during the scope of the reaction. For this purpose, the different reactions were carried out under the optimized conditions using ChCl/glycerol as solvent (Table 3). Regarding the electrophile, an important steric effect was observed, being compound **4** obtained with the best enantioselectivity when using di-*tert*-butyl azodicarboxylate (DBAB) as electrophile (Table 3, entry 3). This electrophile was used for further studies.

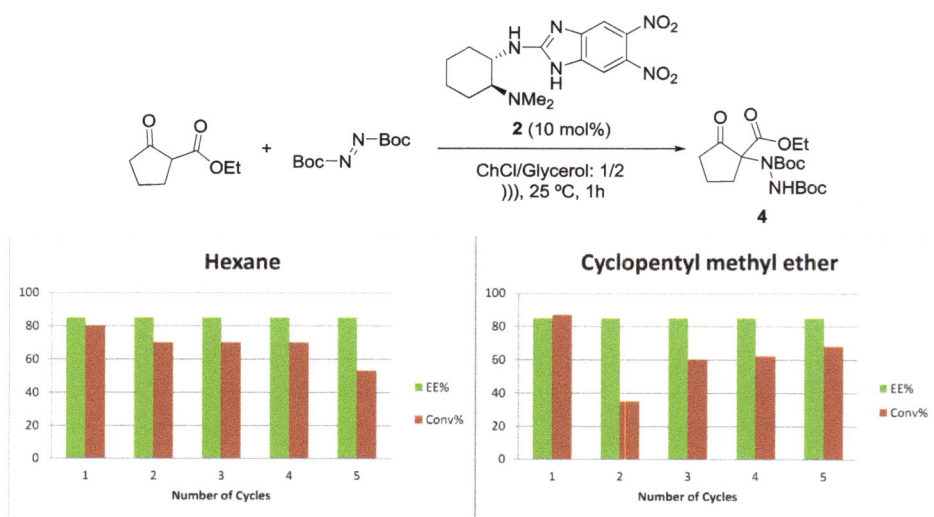

**Scheme 2.** Recycling studies.

The α-amination of other β-ketoesters such as, ethyl 1-oxo-2,3-dihydro-1*H*-indene-2-carboxylate, methyl 1-oxo-2,3-dihydro-1*H*-indene-2-carboxylate, methyl 1-oxo-1,2,3,4-tetrahydronaphthalene-2-carboxylate, and 3-acetyldihydrofuran-2(3*H*)-one was also assessed (Table 3, entries 5–7). In general, good isolated yields were obtained with low enantioselectivites (13 to 36% ee). Better enantioselection was observed in the α-amination with DBAB of 1,3-diketones, especially in the case of 2-acetylcyclopentan-1-one (Table 3, entry 9), which afforded compound **14** in a 75% isolated yield and 53% enantiomeric excess.

Due to accessibility, as well as green considerations, enantioselective organocatalysis has proved to be one of the most efficient approach towards the synthesis of drugs and natural products [39–43]. In particular, the organocatalytic functionalization of indolin-3-one has been recently studied since this type of heterocycles are commonly found in an ample range of biologically active natural alkaloids [44–48]. As depicted in Scheme 4 (Equation (a)), the **2**-catalyzed electrophilic α-amination of methyl 1-acetyl-3-oxoindoline-2-carboxylate [49] in ChCl/glycerol (1/2) as solvent under the optimized reaction conditions gave compound **15** in excellent yields and moderate enantioselectivities (**15a**: R = *i*Pr, 30% ee; **15b**: R = *t*Bu, 45% ee).

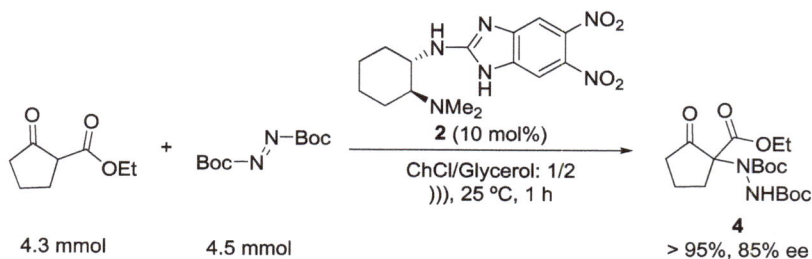

**Scheme 3.** Gram-scale α-amination of ethyl 2-oxocyclopentane-1-carboxylate catalyzed by **2**.

Table 3. Asymmetric α-amination catalyzed by **3**. Reaction scope.

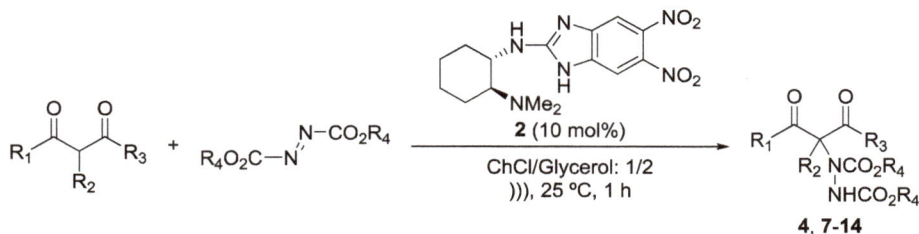

| Entry | Dicarbonyl | Azodicarboxylate | Product | Yield (%) [1] | Ee (%) [2] |
|-------|-----------|------------------|---------|-----------|---------|
| 1 | | BocN=NBoc | 4 | 78 | 85 |
| 2 | | $i$PrO$_2$CN=NCO$_2$$i$Pr | 7 | 52 | 60 |
| 3 | | EtO$_2$CN=NCO$_2$Et | 8 | 76 | 65 |
| 4 | | BnO$_2$CN=NCO$_2$Bn | 9 | 0 | nd |
| 5 | | BocN=NBoc | 10 | 66 | 36 |
| 6 | | BocN=NBoc | 11 | 65 | 35 |
| 7 | | BocN=NBoc | 12 | 65 | 13 |
| 8 | | BocN=NBoc | 13 | 68 | 25 |
| 9 | | BocN=NBoc | 14 | 75 | 53 |

[1] Isolated yield after flash chromatography. [2] Enantiomeric excess determined by chiral HPLC analysis.

On the other hand, 2,2-disubstituted oxindole **16**, which is a precursor of biologically active molecules containing indolin-3-ones with a quaternary stereocenter at the 2-position, such as Brevianamide A, Austamide, among others, has been prepared with excellent yield and diastereoselectivity and a 57% ee by the **2**-catalyzed conjugate addition of methyl 1-acetyl-3-oxoindoline-2-carboxylate to β-nitrostyrene [50] (Scheme 4, Equation (b)). The use of Takemoto's thiourea type catalyst for this transformation led to compound 16 with similar results (>95%, dr > 20:1, 63% ee).

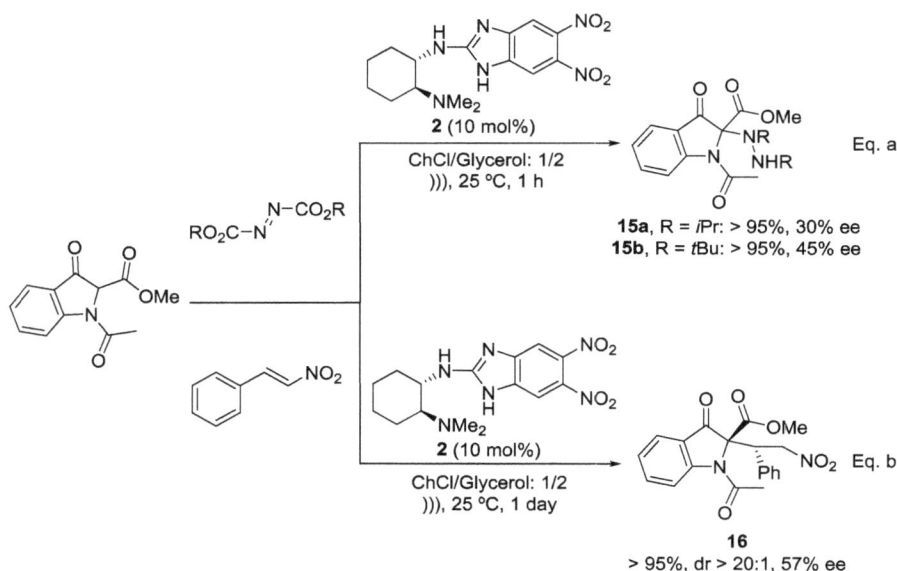

**Scheme 4.** Asymmetric organocatalyzed functionalization of methyl 1-acetyl-3-indol-2-carboxylate in DES.

## 3. Materials and Methods

### 3.1. General

Unless otherwise noted, all commercial reagents and solvents were used without further purification. Reactions under argon atmosphere were carried out in oven-dried glassware sealed with a rubber septum using anhydrous solvents. Melting points were determined with a hot plate apparatus and are uncorrected. [1]H-NMR (300 or 400 MHz) and [13]C-NMR (75 or 101 MHz) spectra were obtained on a Bruker AC-300 or AC-400(Bruke Corporation, Villerica, MA., USA), using CDCl$_3$ as solvent and tetramethyl silane (TMS) (0.003%) as reference, unless otherwise stated. Chemical shifts ($\delta$) are reported in ppm values relative to TMS and coupling constants ($J$) in Hz. Low-resolution mass spectra (MS) were recorded in the electron impact mode (EI, 70 eV, He as carrier phase) using an Agilent 5973 Network Mass Selective Detector spectrometer (Agilent Technologies, Santa Clara, CA, USA), being the samples introduced through a GC chromatograph Agilent 6890N (Agilent Technologies, Santa Clara, CA, USA) equipped with a HP-5MS column [(5%-phenyl)-methylpolysiloxane; length 30 m; ID 0.25 mm; film 0.25 mm]. IR spectra were obtained using a JASCO FT/IR 4100 spectrophotometer (Jasco Analytical Spain, Madrid, Spain) equipped with an ATR component; wavenumbers are given in cm$^{-1}$. Analytical TLC was performed on Merck aluminium sheets with silica gel 60 F254. Analytical TLC was visualized with UV light at 254 nm Silica gel 60 (0.04–0.06 mm) was employed for flash chromatography whereas P/UV254 silica gel with CaSO$_4$ (28–32%) supported on glass plates was employed for preparative TLC. Chiral High-performance liquid chromatography (HPLC) analyses were performed on an Agilent 1100 Series (Agilent Technologies, Santa Clara, CA, USA), (Quat Pump G1311A, DAD G1315B detector and automatic injector) equipped with chiral columns using mixtures of hexane/isopropanol as mobile phase, at 25 °C. The asymmetric reactions were sonicated in an ultrasounds P-Selecta instrument at 360 W.

### 3.2. Synthesis of Catalyst 2

Catalyst **1** [51] (50 mg, 0.2 mmol, 1 equiv.) was dissolved in concentrated $H_2SO_4$ (0.2 mL, 98%) and stirred vigorously for 5 minutes; concentrated $HNO_3$ (0.4 mL, 65%) was then carefully added to the mixture at $-20$ °C. Then, the reaction was stirred at room temperature for 16 h. After this period, the mixture was treated with cold water and basified until pH 8 with a 25% aqueous solution of $NH_3$. Finally, the aqueous phase was extracted with AcOEt (3 × 20 mL). The collected organic phases were dried over anhydrous $MgSO_4$. After filtration, the organic solvent was removed under reduced pressure to give catalyst **2** without further purification as a red solid (74% yield, 52 mg, 0.15 mmol); mp 110–115 °C ($CH_2Cl_2$, decomposes); $\delta_H$ (300 MHz, $CDCl_3$) 1.19–1.49 (m, 4H, 2 × $CH_2$), 1.63–1.98 (m, 4H, 2 × $CH_2$), 2.37 (s, 6H, 2 × Me), 2.51 (m, 1H, CHNMe$_2$), 3.66 (bs, 1H, CHNH),7.49 (s, 2H, ArH); $\delta_C$ (75 MHz, $CDCl_3$) 21.7, 24.4, 24.6, 33.2, 39.8, 53.8, 67.8, 108.3, 136.8, 142.0, 161.8; *m/z* 348 [M$^+$, <1%] 128 (10), 126 (11), 125 (100), 124 (25), 84 (64), 71 (24), 58 (20), 44 (10).

### 3.3. Typical Procedure for the α-Amination Reaction

Catalyst **2** (5.22 mg, 0.015 mmol, 10 mol %) and ethyl 2-oxocyclopentane-1-carboxylate (23.4 mg, 0.15 mmol) were dissolved in a mixture of ChCl/Gly (1/2 molar ratio, 0.2 mL) and kept under stirring for 10 minutes at rt., followed by the addition of di-*tert*-butylazodicarboxylate (36.8 mg, 0.16 mmol). The reaction was vigorously stirred in ultrasounds for 1 h. After this period, water (3 mL) was added to the mixture and the reaction product was extracted with EtOAc (3 × 5 mL). The collected organic phases were dried over anhydrous $MgSO_4$ and, after filtration, the solvent was evaporated under reduced pressure to give crude **4**. Purification by flash column chromatography on silica gel (hexane/EtOAc: 7/3) afforded pure **4** (45.1 mg, 78% yield). $\delta_H$ (300 MHz, $CDCl_3$) 1.28 (t, *J* = 7.1 Hz, 3H), 1.59–1.36 (m, 18H), 2.98–1.75 (m, 6H), 4.24 (m, 2H), 6.53 (br s, 1H) ppm. The enantiomeric excess of **4** was determined by chiral HPLC analysis (Chiralpack IA, hexane/EtOH: 96/04, 0.7 mL/min).

### 3.4. Typical Procedure for the Recovery of the Catalyst in the α-Amination Reaction

A mixture of catalyst **2** (5.22 mg, 0.015 mmol, 10 mol %) and ethyl 2-oxocyclopentane-1-carboxylate (23.4 mg, 0.15 mmol) in ChCl/Gly (1/2 molar ratio, 0.2 mL) was stirred for 10 minutes at rt. Then, di-*tert*-butylazodicarboxylate (36.8 mg, 0.16 mmol) was added. The reaction was vigorously stirred in ultrasounds for 1 h. After this period, the corresponding organic solvent was added (3 mL) and the mixture was stirred for 10 minutes at rt. The stirring was then stopped to allow phase separation and the upper organic layer was removed. This extractive procedure was repeated two more times and the combined organic extracts were washed with water (3 × 5 mL), dried ($MgSO_4$), filtered, and evaporated under reduced pressure. Then, the next reaction cycle was performed with the obtained DES/**2** mixture, adding fresh ethyl 2-oxocyclopentane-1-carboxylate and di-*tert*-butylazodicarboxylate. This reaction mixture was subjected again to the above-described procedure and further reaction cycles were repeated using the recycled deep eutectic solvent phase.

## 4. Conclusions

The enantioselective electrophilic α-amination of 1,3-dicarbonyl compounds with diazodicarboxylates catalyzed by the bifunctional chiral 2-aminobenzimidazole-derivative **2** has been carried out in choline chloride/glycerol or choline chloride/urea deep eutectic solvents. The protocols presented are simple, cheap, clean and scalable. Moreover, the recovery and reuse of the catalyst and reaction medium can be performed at least five times, achieving high and similar enantioselectivities. The synthesis of two natural product precursors is possible by the application of this procedure, as well as the conjugate addition to β-nitrostyrene. With these results, it has been shown that the combination of organocatalyzed enantioselective organic processes in deep eutectic solvents as a reaction media are a clear example of a green, bio-renewable and sustainable process.

**Supplementary Materials:** The following are available online at http://www.mdpi.com/2073-4344/8/5/217/s1.

**Author Contributions:** D.R.Ñ. and P.K. performed the synthetic works. G.G. and D.A.A. designed the experiments of the project and supervised the whole studies reported in the manuscript. D.R.Ñ., P.K., G.G. and D.A.A. wrote the manuscript.

**Acknowledgments:** Financial support from the University of Alicante (UAUSTI16-03, UAUSTI16-10, VIGROB-173), the Spanish Ministerio de Economía, Industria y Competitividad (CTQ2015-66624-P) is acknowledged.

**Conflicts of Interest:** The authors declare no conflicts of interest.

## References and Note

1. Mahrwald, R. *Enantioselective Organocatalyzed Reactions*; Springer: Dordrecht, The Netherlands, 2011.
2. Berkessel, A.; Gröger, H. *Asymmetric Organocatalysis—From Biomimetic Concepts to Applications in Asymmetric Synthesis*; Wiley-VCH: Weinheim, Germany, 2006.
3. Dalko, P. *Enantioselective Organocatalysis*; Wiley-VCH: Weinheim, Germany, 2007.
4. Clarke, C.J.; Tu, W.C.; Levers, O.; Bröhl, A.; Hallett, J.P. Green and Sustainable Solvents in Chemical Processes. *Chem. Rev.* **2018**, *118*, 747–800. [CrossRef] [PubMed]
5. Kerton, F.M. Solvent Systems for Sustainable Chemistry. In *Encyclopedia of Inorganic and Bioinorganic Chemistry*; John Wiley & Sons: New York, NY, USA, 2016.
6. Liu, Y.; Friesen, J.B.; McAlpine, J.B.; Lankin, D.C.; Chen, S.-N.; Pauli, G.F. Natural Deep Eutectic Solvents: Properties, Applications, and Perspectives. *J. Nat. Prod.* **2018**, *81*, 679–690. [CrossRef] [PubMed]
7. Zhang, Q.; De Oliveira Vigier, K.; Royera, S.; Jerome, F. Deep eutectic solvents: Syntheses, properties and applications. *Chem. Soc. Rev.* **2012**, *41*, 7108–7146. [CrossRef] [PubMed]
8. García-Álvarez, J. Deep eutectic solvents and their applications as new green and biorenewable reaction media. In *Handbook of Solvents*; Wypych, G., Ed.; ChemTec Publishing: Toronto, ON, Canada, 2014; Volume 2, pp. 813–844.
9. Guajardo, N.; Müller, C.R.; Schrebler, R.; Carlesi, C.; Domínguez de María, P. Deep eutectic solvents for organocatalysis, biotransformations, and multistep organocatalyst/enzyme combinations. *ChemCatChem* **2016**, *8*, 1020–1027. [CrossRef]
10. Alonso, D.A.; Baeza, A.; Chinchilla, R.; Guillena, G.; Pastor, I.P.; Ramón, D.J. Deep Eutectic Solvents: The Organic Reaction Medium of the Century. *Eur. J. Org. Chem.* **2016**, *2016*, 612–632. [CrossRef]
11. García-Álvarez, J. Deep Eutectic Mixtures: Promising Sustainable Solvents for Metal-Catalysed and Metal-Mediated Organic Reactions. *Eur. J. Inorg. Chem.* **2015**, *2015*, 5147–5157. [CrossRef]
12. García-Álvarez, J.; Hevia, E.; Capriati, V. Reactivity of Polar Organometallic Compounds in Unconventional Reaction Media: Challenges and Opportunities. *Eur. J. Org. Chem.* **2015**, *2015*, 6779–6799. [CrossRef]
13. Massolo, E.; Palmieri, S.; Benaglia, M.; Capriati, V.; Perna, F.M. Stereoselective organocatalysed reactions in deep eutectic solvents: Highly tunable and biorenewable reaction media for sustainable organic synthesis. *Green Chem.* **2016**, *18*, 792–797. [CrossRef]
14. Branco, L.C.; Faisca Phillips, A.M.; Marques, M.M.; Gago, S.; Branco, P.S. Recent advances in sustainable organocatalysis. In *Recent Advances in Organocatalysis*; Karame, I., Srour, H., Eds.; InTech: Rijeka, Croatia, 2016; pp. 141–182. ISBN 978-953-51-2673-7.
15. Genet, J.-P.; Creck, C.; Lavergne, D. *Modern Amination Methods*; Ricci, A., Ed.; Wiley-VCH: Weinheim, Germany, 2000; Chapter 3.
16. *Amino Group Chemistry: From Synthesis to the Life Sciences*; Ricci, A., Ed.; Wiley-VCH: Weinheim, Germany, 2008.
17. Janey, J.M. Recent Advances in Catalytic, Enantioselective α-Aminations and α-Oxygenations of Carbonyl Compounds. *Angew. Chem. Int. Ed.* **2005**, *44*, 4292–4300. [CrossRef] [PubMed]
18. Smith, A.M.R.; Hii, K.K. Transition Metal Catalyzed Enantioselective α-Heterofunctionalization of Carbonyl Compounds. *Chem. Rev.* **2011**, *111*, 1637–1656. [CrossRef] [PubMed]
19. Zhou, F.; Liao, F.-M.; Yu, J.-S.; Zhou, J. Catalytic Asymmetric Electrophilic Amination Reactions to Form Nitrogen-Bearing Tetrasubstituted Carbon Stereocenters. *Synthesis* **2014**, *46*, 2983–3003. [CrossRef]
20. Khose, V.N.; John, M.E.; Pandey, A.D.; Karnik, A.V. Chiral benzimidazoles and their applications in stereodiscrimination processes. *Tetrahedron Asymmetry* **2017**, *28*, 1233–1289. [CrossRef]

21. Nájera, C.; Yus, M. Chiral benzimidazoles as hydrogen bonding organocatalysts. *Tetrahedron Lett.* **2015**, *56*, 2623–2633. [CrossRef]

22. Almaşi, D.; Alonso, D.A.; Gómez-Bengoa, E.; Nájera, C. Chiral 2-Aminobenzimidazoles as Recoverable Organocatalysts for the Addition of 1,3-Dicarbonyl Compounds to Nitroalkenes. *J. Org. Chem.* **2009**, *74*, 6163–6168. [CrossRef] [PubMed]

23. Gómez-Torres, E.; Alonso, D.A.; Gómez-Bengoa, E.; Nájera, C. Conjugate Addition of 1,3-Dicarbonyl Compounds to Maleimides Using a Chiral C2-Symmetric Bis(2-aminobenzimidazole) as Recyclable Organocatalyst. *Org. Lett.* **2011**, *13*, 6106–6109. [CrossRef] [PubMed]

24. Gómez-Torres, E.; Alonso, D.A.; Gómez-Bengoa, E.; Nájera, C. Enantioselective Synthesis of Succinimides by Michael Addition of 1,3-Dicarbonyl Compounds to Maleimides Catalyzed by a Chiral Bis(2-aminobenzimidazole) Organocatalyst. *Eur. J. Org. Chem.* **2013**, *2013*, 1434–1440. [CrossRef]

25. Gómez-Martínez, M.; Alonso, D.A.; Pastor, I.P.; Guillena, G.; Baeza, A. Organocatalyzed Assembly of Chlorinated Quaternary Stereogenic Centers. *Asian J. Org. Chem.* **2016**, *5*, 1428–1437. [CrossRef]

26. Sánchez, D.; Baeza, A.; Alonso, D. Organocatalytic Asymmetric α-Chlorination of 1,3-Dicarbonyl Compounds Catalyzed by 2-Aminobenzimidazole Derivatives. *Symmetry* **2016**, *8*, 3. [CrossRef]

27. Trillo, P.; Gómez-Martínez, M.; Alonso, D.A.; Baeza, A. 2-Aminobenzimidazole Organocatalyzed Asymmetric Amination of Cyclic 1,3-Dicarbonyl Compounds. *Synlett* **2015**, *26*, 95–100.

28. Benavent, L.; Puccetti, F.; Baeza, A.; Gómez-Martínez, M. Readily Available Chiral Benzimidazoles-Derived Guanidines as Organocatalysts in the Asymmetric α-Amination of 1,3-Dicarbonyl Compounds. *Molecules* **2017**, *22*, 1333. [CrossRef] [PubMed]

29. Ñíguez, D.R.; Guillena, G.; Alonso, D.A. Chiral 2-Aminobenzimidazoles in Deep Eutectic Mixtures: Recyclable Organocatalysts for the Enantioselective Michael Addition of 1,3-Dicarbonyl Compounds to β-Nitroalkenes. *ACS Sustain. Chem. Eng.* **2017**, *5*, 10649–10656. [CrossRef]

30. Muller, C.R.; Meiners, I.; Domínguez de María, P. Highly enantioselective tandem enzyme–organocatalyst crossed aldol reactions with acetaldehyde in deep-eutectic-solvents. *RSC Adv.* **2014**, *4*, 46097–46101. [CrossRef]

31. Muller, C.R.; Rosen, A.; Domínguez de María, P. Multi-step enzyme-organocatalyst C–C bond forming reactions in deep-eutectic-solvents: Towards improved performances by organocatalyst design. *Sustain. Chem. Process* **2015**, *3*, 12–20. [CrossRef]

32. Martínez, R.; Berbegal, L.; Guillena, G.; Ramón, D.J. Bio-renewable enantioselective aldol reaction in natural deep eutectic solvents. *Green Chem.* **2016**, *18*, 1724–1730. [CrossRef]

33. Fanjul-Mosteirín, N.; Concellón, C.; del Amo, V. L-Isoleucine in a choline chloride/ethylene glycol deep eutectic solvent: A reusable reaction kit for the asymmetric cross-aldol carboligation. *Org. Lett.* **2016**, *18*, 4266–4269. [CrossRef] [PubMed]

34. Brenna, D.; Massolo, E.; Puglisi, A.; Rossi, S.; Celentano, G.; Benaglia, M.; Capriati, V. Towards the development of continuous, organocatalytic, and stereoselective reactions in deep eutectic solvents. *Beilstein J. Org. Chem.* **2016**, *12*, 2620–2626. [CrossRef] [PubMed]

35. Flores-Ferrándiz, J.; Chinchilla, R. Organocatalytic enantioselective conjugate addition of aldehydes to maleimides in deep eutectic solvents. *Tetrahedron Asymmetry* **2017**, *28*, 302–306. [CrossRef]

36. Mason, T.J. Ultrasound in synthetic organic chemistry. *Chem. Soc. Rev.* **1997**, *26*, 443–451. [CrossRef]

37. Chatel, G.; MacFarlane, D.R. Ionic liquids and ultrasound in combination: Synergies and challenges. *Chem. Soc. Rev.* **2014**, *43*, 8132–8149. [CrossRef] [PubMed]

38. For the full catalyst optimization study, see the SI.

39. Merad, J.; Lalli, C.; Bernadat, G.; Maury, J.; Masson, G. Enantioselective Brønsted Acid Catalysis as a Tool for the Synthesis of Natural Products and Pharmaceuticals. *Chem. Eur. J.* **2018**, *24*, 3925–3943. [CrossRef] [PubMed]

40. Dibello, E.; Gamenara, D.; Seoane, G. Organocatalysis in the Synthesis of Natural Products: Recent Developments in Aldol and Mannich Reactions, and 1,4-Conjugated Additions. *Curr. Organocatal.* **2015**, *2*, 124–149. [CrossRef]

41. Sun, B.-F. Total synthesis of natural and pharmaceutical products powered by organocatalytic reactions. *Tetrahedron Lett.* **2015**, 2133–2140. [CrossRef]

42. Yu, J.; Zhou, Y.; Chen, D.F.; Gong, L.-Z. Organocatalytic asymmetric synthesis of chiral nitrogenous heterocycles and natural products. *Pure Appl. Chem.* **2014**, *86*, 1217–1226. [CrossRef]

43. Abbasov, M.E.; Romo, D. The ever-expanding role of asymmetric covalent organocatalysis in scalable, natural product synthesis. *Nat. Prod. Rep.* **2014**, *31*, 1318–1327. [CrossRef] [PubMed]
44. Yu, B.; Xing, H.; Yu, D.-Q.; Liu, H.-M. Catalytic asymmetric synthesis of biologically important 3-hydroxyoxindoles: An update. *Beilstein J. Org. Chem.* **2016**, *12*, 1000–1039. [CrossRef] [PubMed]
45. Macaev, F.Z.; Sucman, N.S.; Boldescu, V.V. Selective transformations of isatins to substituted 2-oxindoles. *Russ. Chem. Bull.* **2014**, *63*, 15–25. [CrossRef]
46. Dalpozzo, R.; Bartoli, G.; Bencivenni, G. Recent advances in organocatalytic methods for the synthesis of disubstituted 2- and 3-indolinones. *Chem. Soc. Rev.* **2012**, *41*, 7247–7290. [CrossRef] [PubMed]
47. Badillo, J.J.; Hanhan, N.V.; Franz, A.K. Enantioselective synthesis of substituted oxindoles and spirooxindoles with applications in drug discovery. *Curr. Opin. Drug. Discov. Dev.* **2010**, *13*, 758–776.
48. Chauhan, P.; Chimni, S.S. Organocatalytic asymmetric synthesis of 3-amino-2-oxindole derivatives bearing a tetra-substituted stereocenter. *Tetrahedron Asymmetry* **2013**, *24*, 343–356. [CrossRef]
49. Yarlagadda, S.; Ramesh, B.; Reddy, C.R.; Srinivas, L.; Sridhar, B.; Subba Reddy, B.V. Organocatalytic Enantioselective Amination of 2-Substituted Indolin-3-ones: A Strategy for the Synthesis of Chiral α-Hydrazino Esters. *Org. Lett.* **2017**, *19*, 170–173. [CrossRef] [PubMed]
50. Jin, C.-Y.; Wang, Y.; Liu, Y.-Z.; Shen, C.; Xu, P.-F. Organocatalytic Asymmetric Michael Addition of Oxindoles to Nitroolefins for the Synthesis of 2,2-Disubstituted Oxindoles Bearing Adjacent Quaternary and Tertiary Stereocenters. *J. Org. Chem.* **2012**, *77*, 11307–11312. [CrossRef] [PubMed]
51. For the synthesis of catalyst **1** see SI.

**catalysts**

**MDPI**

*Communication*

# Catalytic Enantioselective Addition of Me₂Zn to Isatins

**Carlos Vila \*, Andrés del Campo, Gonzalo Blay and José R. Pedro \***

Departament de Química Orgànica, Facultat de Química, Universitat de València, Dr. Moliner 50,
Burjassot 46100 (València), Spain; andelcam@alumni.uv.es (A.d.C.); gonzalo.blay@uv.es (G.B.)
\*  Correspondence: carlos.vila@uv.es (C.V.); jose.r.pedro@uv.es (J.R.P.);
   Tel.: +34-9635-44510 (C.V.); +34-9635-44329 (J.R.P.)

Received: 15 November 2017; Accepted: 8 December 2017; Published: 13 December 2017

**Abstract:** Chiral α-hydroxyamide **L5** derived from (S)-(+)-mandelic acid catalyzes the enantioselective addition of dimethylzinc to isatins affording the corresponding chiral 3-hydroxy-3-methyl-2-oxindoles with good yields and er up to 90:10. Furthermore, several chemical transformations were performed with the 3-hydroxy-2-oxindoles obtained.

**Keywords:** asymmetric catalysis; isatin; 3-hydroxyoxindole; zinc; mandelamides; chiral α-hydroxyamide

## 1. Introduction

3-Substituted-3-hydroxy-2-oxindole are an important class of compounds that have shown a broad range of biological activities. This scaffold is present in a large variety of natural and synthetic compounds that exhibit pharmaceutical properties [1–8]. Structure–activity relationship studies have shown that the biological activities of these compounds are significantly affected both by the configuration of the C3 and its substitution pattern [9–11]. Therefore, in the last years, the asymmetric synthesis of chiral 3-substituted-3-hydroxy-2-oxindoles have become a hot topic in organic synthesis [12,13]. The synthesis includes allylation [14,15], crotylation [16], arylation [17,18] and decarboxylative cyanomethylation [19] of isatines, as well as the palladium catalyzed intramolecular arylation [20]. The particular interest is the 3-hydroxy-3-methyl-2-oxindole structure, which is present in several natural products such as convolutamydine C [21] and synthetic compounds with biological activities or drug candidates such as compound **2a** [22], compound A [23] and compound B [24] (Figure 1).

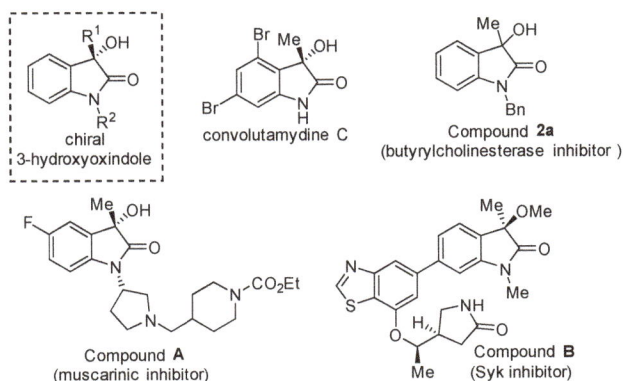

**Figure 1.** Biologically active 3-hydroxy-3-methyl-2-oxindole compounds.

There are few methodologies for the synthesis of chiral 3-hydroxy-3-methyl-2-oxindoles in the literature, and the number of catalytic enantioselective examples is scarce. For example, the asymmetric oxidation of 3-methylindolin-2-one has been described for the synthesis of such compounds [25–27]. However, the most direct and versatile methodology is the enantioselective nucleophilic addition of organometallic reagents to isatins (Scheme 1). In this context, the addition of dialkylzinc reagents to isatins represents an attractive procedure for this purpose [28–33]. Nevertheless, only the group of Shibashaki [34] described just one example of the enantioselective addition of $Me_2Zn$ catalyzed by a proline-derived aminodiol ligand, obtaining the corresponding 3-hydroxy-3-methyl-2-oxindole in 82% yield and 88:12 enantiomeric ratio. In view of this lack of methodologies for the synthesis of such compounds, we decide to study the asymmetric addition of $Me_2Zn$ to isatins catalyzed by $\alpha$-hydroxyamides derived from (S)-(+)-mandelic acid as chiral ligands [35–40].

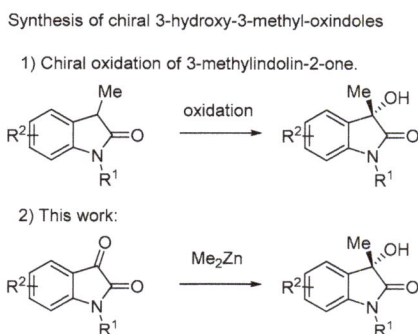

Synthesis of chiral 3-hydroxy-3-methyl-oxindoles

1) Chiral oxidation of 3-methylindolin-2-one.

2) This work:

**Scheme 1.** Asymmetric methodologies for the synthesis of 3-hydroxy-3-methyl-2-oxindole compounds.

## 2. Results

We initiated our studies by evaluating on the addition of $Me_2Zn$ to *N*-benzylisatine (**1a**) in the presence of a series of chiral $\alpha$-hydroxyamides derived from (S)-(+)-mandelic acid as ligands. A 1.2 M $Me_2Zn$ solution in toluene (7 eq.) was added dropwise to a solution of ligand **L1** (0.2 eq.) in 1 mL of toluene at room temperature. After 30 min, a solution of *N*-benzylisatine (**1a**) in 1 mL of toluene was added and the mixture was stirred for 1 h. The corresponding (S)-1-benzyl-3-hydroxy-3-methylindolin-2-one (**2a**) was obtained in 87% yield with 77.5:22.5 enantiomeric ratio (entry 1, Table 1). After, different solvents such as $CH_2Cl_2$, $ClCH_2CH_2Cl$, THF and $Et_2O$ were tested (entries 2–5, Table 1). When $CH_2Cl_2$ and $Et_2O$ were used as solvent, the corresponding product **2a** was obtained with higher enantiomeric ratio, while coordinating solvents such as THF have a detrimental effect in both conversion and enantioselectivity of the reaction (entry 4, Table 1). Therefore, we decided to continue the optimization process with $CH_2Cl_2$ due to solubility problems of the starting material in $Et_2O$. With the best solvent, different $\alpha$-hydroxyamides (Figure 1) were tested as chiral ligands (entries 6–15, Table 1). First, we evaluated the influence of group attached to the nitrogen atom of the amide (Bn, Ph or *t*Bu, entries 1, 6 and 7), obtaining the best enantioselectivity with ligand **L1**. Then, the influence of the substituent in the chiral center of the ligand was evaluated (entry 8). With the corresponding $\alpha$-hydroxy-*N*-benzylamide **L4** derived from (S)-3-phenyllactic acid, product **2a** was afforded with lower er of 75:25. Therefore, we continue the optimization process with $\alpha$-hydroxiamides derived from (S)-(+)-mandelic acid (**L5–L11**). We evaluated the influence of the presence of different groups in the aromatic ring of the amide. Ligand **L5**, prepared from (S)-(+)-mandelic acid and 4-chlorobenzylamine gave the best enantioselectivity on the reaction, obtaining the chiral alcohol with 95% yield and 85:15 er (entry 9). The introduction of an additional methyl group in the benzylic position of the group attached to the nitrogen atom of the amide (entries 14 and 15) had a slightly deleterious effect on the enantioselectivity of the reaction.

**Table 1.** Optimization of the reaction conditions.

| Entry [a] | Ligand (20 mol%) | Solvent | Yield (%) [b] | er [c] |
|---|---|---|---|---|
| 1 | L1 | toluene | 87 | 77.5:22.5 |
| 2 | L1 | CH$_2$Cl$_2$ | 90 | 82:18 |
| 3 | L1 | ClCH$_2$CH$_2$Cl | 78 | 74:26 |
| 4 | L1 | THF | 44 | 61.5:38.5 |
| 5 | L1 | Et$_2$O | 71 | 82.5:17.5 |
| 6 | L2 | CH$_2$Cl$_2$ | 87 | 70.5:29.5 |
| 7 | L3 | CH$_2$Cl$_2$ | 99 | 57:43 |
| 8 | L4 | CH$_2$Cl$_2$ | 88 | 75:25 |
| 9 | L5 | CH$_2$Cl$_2$ | 95 | 85:15 |
| 10 | L6 | CH$_2$Cl$_2$ | 92 | 83:17 |
| 11 | L7 | CH$_2$Cl$_2$ | 71 | 82:18 |
| 12 | L8 | CH$_2$Cl$_2$ | 77 | 74:26 |
| 13 | L9 | CH$_2$Cl$_2$ | 86 | 60.5:39.5 |
| 14 | L10 | CH$_2$Cl$_2$ | 84 | 74:26 |
| 15 | L11 | CH$_2$Cl$_2$ | 99 | 80:20 |

[a] Reaction conditions: 0.1 mmol **1a**, 1.2 M Me$_2$Zn in toluene (0.7 mmol), and ligand in dry solvent (2 mL) at rt for 1 h. [b] Isolated yield after column chromatography. [c] Enantiomeric ratio determined by chiral HPLC.

Consequently, **L5** was chosen for further optimization (Table 2). Lowering the reaction temperature (entries 1–3, Table 2) had a detrimental effect both in yield and enantioselectivity of the reaction. By decreasing the number of the equivalents of Me$_2$Zn, we could improve the enantiomeric ratio to 90:10 in the reaction (entry 6). At this point, we study the effect of the use of additives [34] (entries 7–10) on the enantioselectivity of the reaction. The addition of alcohols had an interesting effect, MeOH inhibits the reaction, while when *i*PrOH or *t*BuOH were added the enantiomeric ratio decreased slightly. Finally, when Ti(O*i*Pr)$_4$ was used as an additive, the corresponding tertiary alcohol **2a** was obtained with very low enantioselectivity (entry 10). Therefore, we decided as optimized reaction conditions the ones presented in entry 6, Table 2.

With the optimized reaction conditions established, the scope of the reaction was explored (see Supplementary Materials). Initially, *N*-substitution of the oxindole nitrogen atom was evaluated. Groups such as benzyl, methyl [41], allyl or propargyl were tolerated (entries 1, 3–5, Table 3), providing the corresponding tertiary alcohols with good enantioselectivities. However, unprotected free NH group on isatin was not tolerated (entry 2, Table 3), and the corresponding product **2b** was obtained with lower yield and enantioselectivity, as well when the protecting group was acetyl (entry 7) or Ts (entry 8).

**Table 2.** Optimization of the reaction conditions.

| Entry [a] | T (°C) | Additive (X mol%) | Yield (%) [b] | er [c] |
|---|---|---|---|---|
| 1 | −20 | - | 67 | 75:25 |
| 2 | 0 | - | 72 | 79.5:20.5 |
| 3 | 10 | - | 86 | 84.5:15.5 |
| 4 | rt | - | 95 | 85:15 |
| 5 [d] | rt | - | 89 | 89:11 |
| 6 [e] | rt | - | 85 | 90:10 |
| 7 [e,f] | rt | MeOH (40 mol%) | - | - |
| 8 [e,g] | rt | *i*PrOH (40 mol%) | 86 | 88:12 |
| 9 [e,g] | rt | *t*BuOH (40 mol%) | 48 | 86.5:13.5 |
| 10 [e,g] | rt | Ti(O*i*Pr)$_4$ (100 mol%) | 51 | 57:43 |

[a] Reaction conditions: 0.1 mmol **1a**, 1.2 M Me$_2$Zn in toluene (0.7 mmol), and **L5** (20 mol%) in CH$_2$Cl$_2$ (2 mL) for 1 h. [b] Isolated yield after column chromatography. [c] Enantiomeric excess determined by chiral HPLC. [d] 0.35 mmol of Me$_2$Zn was used. [e] 0.2 mmol of Me$_2$Zn was used. [f] The reaction time was 24 h. [g] The reaction time was 4 h.

**Table 3.** Evaluation of the protecting group of the isatin.

| Entry [a] | R$^1$ | 1 | t (h) | 2 | Y (%) [b] | er [c] |
|---|---|---|---|---|---|---|
| 1 | Bn- | **1a** | 1 | **2a** | 85 | 90:10 |
| 2 [d] | H | **1b** | 4 | **2b** | 47 | 61:39 |
| 3 | Me | **1c** | 3 | **2c** | 66 | 82:18 |
| 4 | allyl | **1d** | 3 | **2d** | 71 | 87:13 |
| 5 | propargyl | **1e** | 2 | **2e** | 65 | 83.5:16.5 |
| 6 | CH$_2$CO$_2$Me | **1f** | 3 | **2f** | 70 | 72:28 |
| 7 | COMe | **1g** | 2 | **2g** | 45 | 55:45 |
| 8 | Ts | **1h** | 2 | **2h** | 36 | 72:28 |
| 9 | | **1i** | 1 | **2i** | 81 | 87:13 |

[a] Reaction conditions: 0.1 mmol **1**, 1.2 M Me$_2$Zn in toluene (0.2 mmol), and **L5** (20 mol%) in CH$_2$Cl$_2$ (2 mL). [b] Isolated yield after column chromatography. [c] Enantiomeric ratio determined by chiral HPLC. [d] 0.3 mmol of Me$_2$Zn was used.

Next, the effect of substitution in the benzene ring of the *N*-benzyl protected isatins was studied (Scheme 2). A reduction in the catalyst loading to 10 mol% was also investigated, observing similar conversion and enantioselectivity. Different electron-donating (Me or MeO) or electron-withdrawing (F or Cl) in positions 5, 6 and 7, were tolerated and the corresponding chiral tertiary alcohols were obtained with good yields and enantiomeric ratios from 80:20 to 90:10. However, the presence of a strong electron-withdrawing group (NO$_2$) led to a considerable decrease in the enantiomeric ratio of the reaction product.

To evaluate the potential scalability of the asymmetric addition of Me$_2$Zn to isatins, this procedure was also performed on a 1 mmol scale. As shown in Scheme 3, the corresponding product **2a** was isolated in 98% yield and 88:12 enantiomeric ratio (er).

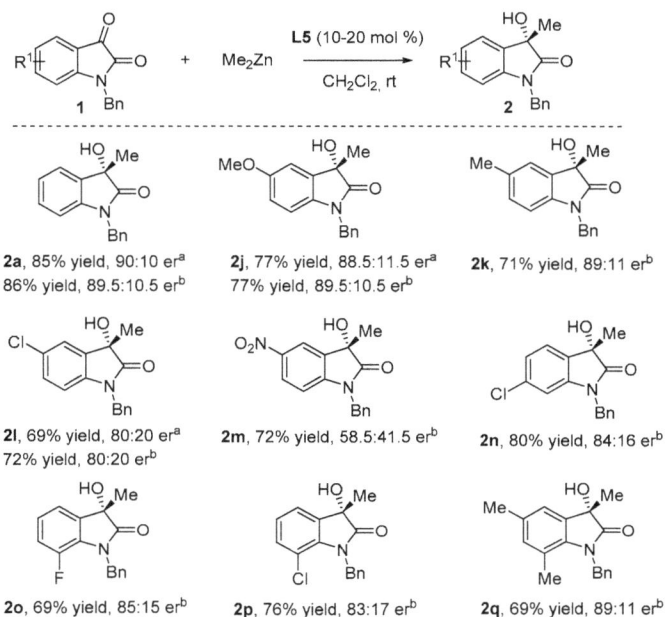

**Scheme 2.** Scope of the enantioselective addition of Me₂Zn to isatins. Reaction conditions: 0.1 mmol **1**, 1.2 M Me₂Zn in toluene (0.2 mmol), and **L5** in CH₂Cl₂ (2 mL). Isolated yield after column chromatography. Enantiomeric ratio determined by chiral HPLC. [a] 20 mol% of **L5** was used. [b] 10 mol% of **L5** was used.

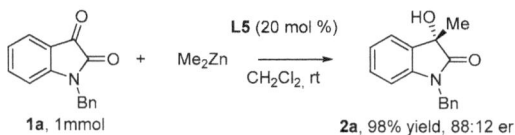

**Scheme 3.** 1 mmol scale reaction. Reaction conditions: 1 mmol **1**, 1.2 M Me₂Zn in toluene (2 mmol), and **L5** (20 mol%) in CH₂Cl₂ (20 mL). Isolated yield after column chromatography. Enantiomeric ratio determined by chiral HPLC.

To highlight the synthetic utility of this methodology, we have applied several chemical transformations (Scheme 4). We tried to reduce the amide moiety of the oxindole **2a** with LiAlH₄, however the epoxide **3a** was obtained. We had some problems to purify epoxide **3a** due to its instability. Nevertheless, we could react compound **3a** with TMSCN, to afford smoothly the corresponding chiral indoline **4a** with 2 stereogenic centers in 65% yield and without losing the enantiomeric purity of compound **2a**.

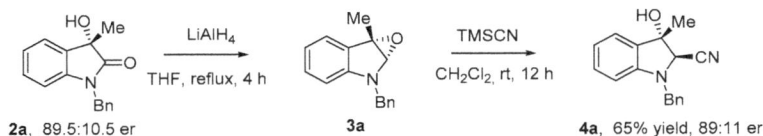

**Scheme 4.** Synthetic transformations of chiral 3-hydroxy-3-methyl-2-oxindole **2a**.

## 3. Materials and Methods

### 3.1. General Information

Reactions were carried out under nitrogen in test tubes or round bottom flasks oven-dried overnight at 120 °C. Dicloromethane, 1,2-dichloroethane and toluene were distilled from $CaH_2$. Tetrahydrofuran (THF) and $Et_2O$ were distilled from sodium benzophenone ketyl. Reactions were monitored by TLC (thin layer chromatography) analysis using Merck Silica Gel 60 F-254 thin layer plates. Flash column chromatography was performed on Merck silica gel 60, 0.040–0.063 mm. Melting points were determined in capillary tubes. NMR (Nuclear Magnetic Resonance) spectra were run in a Bruker DPX300 spectrometer (Bruker, Billerica, MA, USA) at 300 MHz for [1]H and at 75 MHz for [13]C using residual non-deuterated solvent as internal standard ($CHCl_3$: $\delta$ 7.26 and 77.0 ppm). Chemical shifts are given in ppm. The carbon type was determined by DEPT (Distortionless Enhancement by Polarization Transfer) experiments. High resolution mass spectra (ESI) were recorded on a TRIPLETOF[T]5600 spectrometer (AB Sciex, Warrington, UK) equipped with an electrospray source with a capillary voltage of 4.5 kV (ESI). Specific optical rotations were measured using sodium light (D line 589 nm). Chiral HPLC (High performance liquid chromatography) analyses were performed in a chromatograph equipped with a UV diode-array detector using chiral stationary columns from Daicel. 1.2 M $Me_2Zn$ solution in toluene was purchased from Acros (Geel, Belgium). Chiral $\alpha$-hydroxyamides were prepared as described in the literature [35]. Commercially available isatins were used as received. *N*-protected isatins **1** were prepared as described in the literature [42].

### 3.2. Typical Procedures and Characterization Data for Compounds 2

#### 3.2.1. General Procedure for the Enantioselective Addition of $Me_2Zn$ to Isatins

A 1.2 M $Me_2Zn$ solution in toluene (0.17 mL, 0.2 mmol) was added dropwise on a solution of **L5** (5.5 mg, 0.02 mmol or 2.25 mg, 0.01 mmol) in $CH_2Cl_2$ (1 mL) at room temperature under nitrogen. After stirring 30 min, a solution of isatin **1** (0.1 mmol) in $CH_2Cl_2$ (1.0 mL) was added via syringe. The reaction was stirred until the reaction was complete (TLC). The reaction mixture was quenched with $NH_4Cl$ (10 mL), extracted with $CH_2Cl_2$ (3 × 15 mL), washed with brine (10 mL), dried over $MgSO_4$ and concentrated under reduced pressure. Purification by flash chromatography on silica gel afforded compound **2**.

#### 3.2.2. General Procedure for the Non-Enantioselective Addition of $Me_2Zn$ to Isatins

A 1.2 M $Me_2Zn$ solution in toluene (0.17 mL, 0.2 mmol) was added dropwise on a solution of isatin **1** (0.1 mmol) in $CH_2Cl_2$ (2 mL) at room temperature under nitrogen. The reaction was stirred until the reaction was complete (TLC). The reaction mixture was quenched with $NH_4Cl$ (10 mL), extracted with $CH_2Cl_2$ (3 × 15 mL), washed with brine (10 mL), dried over $MgSO_4$ and concentrated under reduced pressure. Purification by flash chromatography on silica gel afforded compound **2**.

*(S)-1-Benzyl-3-hydroxy-3-methylindolin-2-one* (**2a**) [43–45]: Enantiomeric ratio (90:10) was determined by chiral HPLC (Chiralpak AD-H), hexane-iPrOH 80:20, 1 mL/min, major enantiomer rt = 9.3 min, minor enantiomer rt = 8.1 min. White solid; mp = 110–112 °C; $[\alpha]_{20}^D$ = −34.1 (c = 1.09, $CHCl_3$) (90:10 er); [1]H NMR (300 MHz, $CDCl_3$) $\delta$ 7.40 (ddd, *J* = 7.4, 1.2, 0.6 Hz, 1H), 7.34–7.23 (m, 5H), 7.22–7.15 (m, 1H), 7.05 (td, *J* = 7.6, 0.7 Hz, 1H), 6.70 (d, *J* = 7.9 Hz, 1H), 4.94 (d, *J* = 15.7 Hz, 1H), 4.80 (d, *J* = 15.7 Hz, 1H), 2.90 (s, 1H), 1.65 (s, 3H). [13]C NMR (75 MHz, $CDCl_3$) $\delta$ 178.56 (C), 141.91 (C), 135.44 (C), 131.30 (C), 129.53 (CH), 128.83 (CH), 127.70 (CH), 127.18 (CH), 123.49 (CH), 123.24 (CH), 109.56 (CH), 73.69 (C), 43.72 ($CH_2$), 25.08 ($CH_3$); HRMS (ESI) *m/z*: 254.1171 [M + H]$^+$, $C_{16}H_{16}NO_2$ required 254.1176.

*(S)-3-Hydroxy-3-methylindolin-2-one* (**2b**) [46–48]: Enantiomeric ratio (61:39) was determined by chiral HPLC (Chiralpak OD-H), hexane-iPrOH 80:20, 1 mL/min, major enantiomer rt = 5.9 min, minor

enantiomer rt = 7.0 min. White solid; mp = 150–154 °C; $[\alpha]_{20}^{D}$ = −12.84 (c = 0.345, CHCl$_3$) (61:39 er); $^1$H NMR (300 MHz, CDCl$_3$) δ 7.76 (s, 1H), 7.40 (dd, *J* = 7.4, 0.6 Hz, 1H), 7.27 (td, *J* = 7.7, 1.3 Hz, 1H), 7.09 (td, *J* = 7.6, 1.0 Hz, 1H), 6.88 (d, *J* = 7.7 Hz, 1H), 2.82 (s, 1H), 1.62 (s, 3H). $^{13}$C NMR (75 MHz, CDCl$_3$) δ 180.59 (C), 140.11 (C), 132.09 (C), 130.07 (CH), 124.32 (CH), 123.67 (CH), 110.64 (CH), 74.28 (C), 25.25 (CH$_3$).

*(S)-3-Hydroxy-1,3-dimethylindolin-2-one* (**2c**) [35,43,44,49]: Enantiomeric ratio (82:18) was determined by chiral HPLC (Chiralpak AS-H), hexane-iPrOH 90:10, 1.0 mL/min, major enantiomer rt = 15.4 min, minor enantiomer rt = 12.5 min. White solid; mp = 100–104 °C $[\alpha]_{20}^{D}$ = −31.8 (c = 0.59, CHCl$_3$) (82:18 er); $^1$H NMR (300 MHz, CDCl$_3$) δ 7.39 (ddd, *J* = 7.2, 1.3, 0.6 Hz, 1H), 7.30 (td, *J* = 7.7, 1.3 Hz, 1H), 7.08 (td, *J* = 7.5, 1.0 Hz, 1H), 6.82 (dt, *J* = 7.9, 0.8 Hz, 1H), 3.21 (s, 1H), 3.17 (s, 3H), 1.58 (s, 3H). $^{13}$C NMR (75 MHz, CDCl$_3$) δ 178.58 (C), 142.78 (C), 131.43 (C), 129.56 (CH), 123.40 (CH), 123.21 (CH), 108.47 (CH), 73.65 (C), 26.20 (CH$_3$), 24.81 (CH$_3$).

*(S)-1-Allyl-3-hydroxy-3-methylindolin-2-one* (**2d**): Enantiomeric ratio (87:13) was determined by chiral HPLC (Chiralpak AD-H), hexane-iPrOH 80:20, 1.0 mL/min, major enantiomer rt = 6.31 min, minor enantiomer rt = 5.90 min. Oil; $[\alpha]_{20}^{D}$ = −39.2 (c = 0.71, CHCl$_3$) (87:13 er); $^1$H NMR (300 MHz, CDCl$_3$) δ 7.40 (ddd, *J* = 7.4, 1.4, 0.6 Hz, 1H), 7.26 (td, *J* = 7.8, 1.4 Hz, 1H), 7.07 (td, *J* = 7.5, 1.0 Hz, 1H), 6.81 (dd, *J* = 7.9, 0.8 Hz, 1H), 5.81 (ddt, *J* = 17.3, 10.4, 5.3 Hz, 1H), 5.24–5.20 (m, 1H), 5.19–5.15 (m, 1H), 4.34 (ddt, *J* = 16.4, 5.2, 1.7 Hz, 1H), 4.23 (ddt, *J* = 16.4, 5.3, 1.7 Hz, 1H), 3.16 (s, 1H), 1.60 (s, 3H); $^{13}$C NMR (75 MHz, CDCl$_3$) 178.31 (C), 141.95 (C), 131.39 (C), 131.05 (CH), 129.46 (CH), 123.48 (CH), 123.17 (CH), 117.67 (CH$_2$), 109.39 (CH), 73.60 (C), 42.26(CH$_2$), 25.01 (CH$_3$); HRMS (ESI) *m/z*: 204.1013 [M + H]$^+$, C$_{12}$H$_{14}$NO$_2$ required 204.1019.

*(S)-3-Hydroxy-3-methyl-1-(prop-2-yn-1-yl)indolin-2-one* (**2e**): Enantiomeric ratio (83.5:16.5) was determined by chiral HPLC (Chiralpak IC), hexane-iPrOH 90:10, 1.0 mL/min, major enantiomer rt = 21.2 min, minor enantiomer rt = 16.9 min. White solid; mp = 84–86 °C; $[\alpha]_{20}^{D}$ = −25.3 (c = 0.66, CHCl$_3$) (83.5:16.5 er); $^1$H NMR (300 MHz, CDCl$_3$) δ 7.43 (ddd, *J* = 7.4, 1.4, 0.6 Hz, 1H), 7.35 (td, *J* = 7.7, 1.3 Hz, 1H), 7.14 (td, *J* = 7.5, 1.0 Hz, 1H), 7.06 (dt, *J* = 7.8, 0.8 Hz, 1H), 4.53 (dd, *J* = 17.7, 2.5 Hz, 1H), 4.41 (dd, *J* = 17.7, 2.5 Hz, 1H), 3.08 (s, 1H), 2.24 (t, *J* = 2.5 Hz, 1H), 1.61 (s, 3H). $^{13}$C NMR (75 MHz, CDCl$_3$) δ 177.52 (C), 140.86 (C), 131.24 (C), 129.58 (CH), 123.59 (CH), 123.52 (CH), 109.57 (CH), 73.69 (C), 73.66 (C), 72.62 (CH), 29.34 (CH$_2$), 24.81 (CH$_3$); HRMS (ESI) *m/z*: 202.0862 [M + H]$^+$, C$_{12}$H$_{12}$NO$_2$ required 202.0863.

*Methyl 2-(3-hydroxy-3-methyl-2-oxoindolin-1-yl)acetate* (**2f**): Enantiomeric ratio (72:28) was determined by chiral HPLC quiral (Chiralpak IC), hexane-iPrOH 90:10, 1.0 mL/min, major enantiomer rt = 52.8 min, minor enantiomer rt = 57.2 min. Yelow solid; mp = 142–144 °C $[\alpha]_{20}^{D}$ = +1.91 (c = 0.82, CHCl$_3$) (72:28 er); $^1$H NMR (300 MHz, CDCl$_3$) δ 7.35 (dd, *J* = 7.3, 1.3 Hz, 1H), 7.21 (dd, *J* = 7.8, 1.3 Hz, 1H), 7.04 (td, *J* = 7.5, 1.0 Hz, 1H), 6.65 (dd, *J* = 7.8, 0.8 Hz, 1H), 4.44 (d, *J* = 17.6 Hz, 1H), 4.29 (d, *J* = 17.5 Hz, 1H), 3.67 (s, 3H), 3.06 (s, 1H), 1.55 (s, 3H). $^{13}$C NMR (75 MHz, CDCl$_3$) δ 178.37 (C), 167.96 (C), 141.34 (C), 131.22 (C), 129.58 (CH), 128.90 (CH), 123.59 (CH), 108.43 (CH), 73.59 (C), 52.66 (CH$_3$), 41.09 (CH$_2$), 24.84 (CH$_3$); HRMS (ESI) *m/z*: 236.0913 [M + H]$^+$, C$_{12}$H$_{14}$NO$_4$ required 236.0917.

*1-Acetyl-3-hydroxy-3-methylindolin-2-one* (**2g**) [50]: Enantiomeric ratio (54:46) was determined by chiral HPLC (Chiralpak IC), hexane-iPrOH 90:10, 1.0 mL/min, major enantiomer rt = 8.3 min, minor enantiomer rt = 7.2 min. White solid; mp = 109–110 °C; $[\alpha]_{20}^{D}$ = −4.8 (c = 0.465, CHCl$_3$) (54:46 er); $^1$H NMR (300 MHz, CDCl$_3$) δ 8.26–8.20 (m, 1H), 7.46 (ddd, *J* = 7.3, 1.5, 0.6 Hz, 1H), 7.38 (ddd, *J* = 8.3, 7.6, 1.5 Hz, 1H), 7.26 (td, *J* = 7.4, 1.1 Hz, 1H), 2.81 (s, 1H), 2.66 (s, 3H), 1.65 (s, 3H). $^{13}$C NMR (75 MHz, CDCl$_3$) δ 179.04 (C), 170.74 (C), 139.10 (C), 130.33 (C), 130.15 (CH), 125.81 (CH), 123.23 (CH), 116.90 (CH), 73.59 (C), 26.47 (CH$_3$), 25.65 (CH$_3$); HRMS (ESI) *m/z*: 228.0632 [M + Na]$^+$, C$_{11}$H$_{11}$NO$_3$Na required 228.0631.

*3-Hydroxy-3-methyl-1-tosylindolin-2-one* (**2h**): Enantiomeric ratio (72:28) was determined by chiral HPLC (Chiralpak AD-H), hexane-iPrOH 80:20, 1.0 mL/min, major enantiomer rt = 11.7 min, minor enantiomer rt = 13.2 min. White solid; mp = 93–95 °C; $[\alpha]_{20}^{D}$ = +7.07 (c = 0.355, CHCl$_3$) (72:28 er); $^1$H NMR (300 MHz, CDCl$_3$) δ 7.97 (d, *J* = 8.4 Hz, 2H), 7.91 (dd, *J* = 8.6, 1.0 Hz, 1H), 7.44–7.35 (m, 2H), 7.32 (dd, *J* = 8.7, 0.7 Hz, 2H), 7.25–7.18 (m, 1H), 2.56 (s, 1H), 2.41 (s, 3H), 1.56 (s, 3H). $^{13}$C NMR (75 MHz, CDCl$_3$) δ 176.82 (C), 145.89 (C), 138.08 (C), 134.83 (C), 130.36 (CH), 130.16 (C), 129.89 (CH), 127.87 (CH), 127.70 (CH), 125.45 (CH), 113.87 (CH), 73.64 (C), 25.75 (CH$_3$), 21.70 (CH$_3$); HRMS (ESI) *m/z*: 300.0689 [M − H$_2$O]$^+$, C$_{16}$H$_{14}$NO$_3$S required 300.0689.

*(S)-3-Hydroxy-3-methyl-1-(naphthalen-1-ylmethyl)indolin-2-one* (**2i**): Enantiomeric ratio (87:13) was determined by chiral HPLC (Chiralpak AS-H), hexane-iPrOH 80:20, 1.0 mL/min, major enantiomer rt = 14.3 min, minor enantiomer rt = 10.9 min. White solid, mp = 131–133 °C; $[\alpha]_{20}^{D}$ = −19.06 (c = 1.23, CHCl$_3$) (87:13 er); $^1$H NMR (300 MHz, CDCl$_3$) δ 8.12–8.06 (m, 1H), 7.89 (dd, *J* = 8.1, 1.5 Hz, 1H), 7.79 (dt, *J* = 8.2, 1.0 Hz, 1H), 7.64–7.49 (m, 2H), 7.47–7.42 (m, 1H), 7.37 (dd, *J* = 8.2, 7.1 Hz, 1H), 7.28 (dd, *J* = 7.1, 1.2 Hz, 1H), 7.12 (dd, *J* = 7.7, 1.5 Hz, 1H), 7.09–7.02 (m, 1H), 6.68 (dt, *J* = 8.0, 0.9 Hz, 1H), 5.52 (d, *J* = 16.2 Hz, 1H), 5.21 (d, *J* = 16.2 Hz, 1H), 3.32 (s, 1H), 1.72 (s, 3H). $^{13}$C NMR (75 MHz, CDCl$_3$) δ 178.84 (C), 142.14 (C), 133.83 (C), 131.44 (C), 130.97 (C), 130.19 (C), 129.49 (CH), 128.93 (CH), 128.40 (CH), 126.56 (CH), 126.0 (C), 125.25 (CH), 124.52 (CH), 123.44 (CH), 123.27 (CH), 122.75 (CH), 109.95 (CH), 73.78 (C), 41.97 (CH$_2$), 25.19 (CH$_3$); HRMS (ESI) *m/z*: 304.1332 [M + H]$^+$, C$_{20}$H$_{18}$NO$_2$ required 304.1332.

*(S)-1-Benzyl-3-hydroxy-5-methoxy-3-methylindolin-2-one* (**2j**): Enantiomeric ratio (89.5:10.5) was determined by chiral HPLC (Chiralpak AD-H), hexane-iPrOH 80:20, 1.0 mL/min, major enantiomer rt = 14.1 min, minor enantiomer rt = 10.3 min. Oil; $[\alpha]_{20}^{D}$ = −36.51 (c = 1.09, CHCl$_3$) (89.5:10.5 er); $^1$H NMR (300 MHz, CDCl$_3$) δ 7.43–7.17 (m, 5H), 7.04 (d, *J* = 2.6 Hz, 1H), 6.71 (dd, *J* = 8.6, 2.6 Hz, 1H), 6.59 (d, *J* = 8.5 Hz, 1H), 4.92 (d, *J* = 15.6 Hz, 1H), 4.77 (d, *J* = 15.7 Hz, 1H), 3.76 (s, 3H), 3.47 (s, 1H), 1.66 (s, 3H). $^{13}$C NMR (75 MHz, CDCl$_3$) δ 178.54 (C), 156.45 (C), 135.48 (C), 135.02 (C), 132.68 (C), 128.79 (CH), 127.64 (CH), 127.14 (CH), 114.08 (CH), 110.48 (CH), 110.11 (CH), 74.06 (C), 55.76 (CH$_3$), 43.76 (CH$_2$), 25.19 (CH$_3$); HRMS (ESI) *m/z*: 284.1280 [M + H]$^+$, C$_{17}$H$_{18}$NO$_3$ required 284.1281.

*(S)-1-Benzyl-3-hydroxy-3,5-dimethylindolin-2-one* (**2k**) [44]: Enantiomeric ratio (89:11) was determined by chiral HPLC (Chiralpak AD-H), hexane-iPrOH 80:20, 1.0 mL/min, major enantiomer rt = 8.4 min, minor enantiomer rt = 7.0 min. White solid; mp = 131–132 °C; $[\alpha]_{20}^{D}$ = −2.33 (c = 0.81, CHCl$_3$) (89:11 er); $^1$H NMR (300 MHz, CDCl$_3$) δ 7.53 (d, *J* = 2.0 Hz, 1H), 7.37–7.21 (m, 6H), 6.58 (dd, *J* = 8.5, 0.9 Hz, 1H), 4.93 (d, *J* = 15.7 Hz, 1H), 4.80 (d, *J* = 15.7 Hz, 1H), 3.11 (s, 1H), 1.66 (s, 3H). $^{13}$C NMR (75 MHz, CDCl$_3$) δ 178.09 (C), 140.87 (C), 134.92 (C), 133.27 (C), 132.31 (CH), 128.94 (CH), 127.90 (CH), 127.11 (CH), 126.96 (CH), 116.03 (C), 111.11 (CH), 73.67 (C), 43.81 (CH2), 25.08 (CH3); HRMS (ESI) *m/z*: 268.1331 [M + H]$^+$, C$_{17}$H$_{18}$NO$_2$ required 268.1332.

*(S)-1-Benzyl-5-chloro-3-hydroxy-3-methylindolin-2-one* (**2l**): Enantiomeric ratio (80:20) was determined by chiral HPLC (Chiralpak AD-H), hexane-iPrOH 80:20, 1.0 mL/min, major enantiomer rt = 8.8 min, minor enantiomer rt = 6.8 min. White siolid; mp = 159–161 °C; $[\alpha]_{20}^{D}$ = −29.37 (c = 0.985, CHCl$_3$) (80:20 er); $^1$H NMR (300 MHz, CDCl$_3$) δ 7.39 (d, *J* = 2.1 Hz, 1H), 7.34–7.21 (m, 5H), 7.16 (dd, *J* = 8.4, 2.2 Hz, 1H), 6.62 (d, *J* = 8.3 Hz, 1H), 4.93 (d, *J* = 15.7 Hz, 1H), 4.79 (d, *J* = 15.7 Hz, 1H), 3.40 (s, 1H), 2.30 (s, 3H), 1.66 (s, 3H). $^{13}$C NMR (75 MHz, CDCl$_3$) δ 178.35 (C), 140.29 (C), 134.94 (C), 133.08 (C), 133.03 (CH), 129.34 (CH), 128.92 (C), 128.77 (CH), 127.87 (CH), 127.10 (CH), 124.19 (CH), 110.61 (CH), 73.73 (C), 43.82 (CH$_2$), 25.05 (CH$_3$), 20.98 (CH$_3$). HRMS (ESI) *m/z*: 288.0782 [M + H]$^+$, C$_{16}$H$_{15}$ClNO$_2$ required 288.0786.

*(S)-1-Benzyl-3-hydroxy-3-methyl-5-nitroindolin-2-one* (**2m**): Enantiomeric ratio (58.5:41.5) was determined by chiral HPLC (Chiralpak AD-H), hexane-iPrOH 80:20, 1.0 mL/min, major enantiomer rt = 12.8 min, minor enantiomer rt = 10.2 min. Oil; $[\alpha]_{20}^{D}$ = −10.9 (c = 1.07, CHCl$_3$) (58.5:41.5 er); 1H NMR (300 MHz, CDCl$_3$) δ 8.29 (d, *J* = 2.2 Hz, 1H), 8.15 (ddd, *J* = 8.8, 2.4, 0.8 Hz, 1H), 7.42–7.19 (m, 5H), 6.80 (dd, *J* = 8.5, 0.8 Hz, 1H), 4.99 (d, *J* = 15.8 Hz, 1H), 4.88 (d, *J* = 15.7 Hz, 1H), 3.67 (s,1H), 1.72 (s, 3H). $^{13}$C NMR (75 MHz, CDCl$_3$) δ 178.95 (C), 147.41 (C), 143.93 (C), 134.24 (C), 132.27 (C), 129.12 (CH),

128.22 (CH), 127.10 (CH), 126.48 (CH), 119.61 (CH), 109.33 (CH), 73.29 (C), 44.10 (CH$_2$), 24.90 (CH$_3$); HRMS (ESI) *m/z*: 298.1027 [M + H]$^+$, C$_{16}$H$_{15}$N$_2$O$_4$ required 299.1026.

*(S)-1-Benzyl-6-chloro-3-hydroxy-3-methylindolin-2-one* (**2n**): Enantiomeric ratio (84:16) was determined by chiral HPLC (Chiralpak AD-H), hexane-iPrOH 80:20, 1.0 mL/min, major enantiomer rt = 7.3 min, minor enantiomer rt = 6.8 min. White solid; mp = 140–141 °C; $[\alpha]_{20}^{D}$ = −18.3 (c = 1.15, CHCl$_3$) (84:16 er); $^1$H NMR (300 MHz, CDCl$_3$) δ 7.38–7.22 (m, 6H), 7.04 (dd, *J* = 7.9, 1.8 Hz, 1H), 6.71 (d, *J* = 1.7 Hz, 1H), 4.92 (d, *J* = 15.7 Hz, 1H), 4.76 (d, *J* = 15.8 Hz, 1H), 3.36 (s, 1H), 1.65 (s, 3H). $^{13}$C NMR (75 MHz, CDCl$_3$) δ 178.68 (C), 143.08 (C), 135.23 (C), 134.86 (C), 129.77 (C), 128.97 (CH), 127.92 (CH), 127.10 (CH), 124.50 (CH), 123.19 (CH), 110.18 (CH), 73.38 (C), 43.80 (CH$_2$), 25.02 (CH$_3$); HRMS (ESI) *m/z*: 288.0783 [M + H]$^+$, C$_{16}$H$_{15}$ClNO$_2$ required 288.0786.

*(S)-1-Benzyl-7-fluoro-3-hydroxy-3-methylindolin-2-one* (**2o**): Enantiomeric ratio (85:15) was determined by chiral HPLC (Chiralpak AD-H), hexane-iPrOH 80:20, 1.0 mL/min, major enantiomer rt = 7.7 min, minor enantiomer rt = 6.6 min. White solid; mp = 106–108 °C; $[\alpha]_{20}^{D}$ = −20.85 (c = 0.93, CHCl$_3$) (85:15 er); $^1$H NMR (300 MHz, CDCl$_3$) δ 7.34–7.24 (m, 5H), 7.23–7.19 (m, 1H), 7.06–6.93 (m, 2H), 5.05 (d, *J* = 16.6 Hz, 1H), 4.98 (d, *J* = 16.6 Hz, 1H), 3.32 (s, 1H), 1.65 (s, 3H). $^{13}$C NMR (75 MHz, CDCl$_3$) δ 178.47 (C), 147.57 (d, *J*$_{C\text{-}F}$ = 244.9 Hz, C), 136.62 (C), 134.32 (d, *J*$_{C\text{-}F}$ = 2.8 Hz, C), 128.62 (CH), 128.36 (d, *J*$_{C\text{-}F}$ = 8.7 Hz, C), 127.63 (CH), 127.35 (d, *J*$_{C\text{-}F}$ = 1.4 Hz, CH), 124.12 (d, *J*$_{C\text{-}F}$ = 6.4 Hz, CH), 119.39 (d, *J*$_{C\text{-}F}$ = 3.3 Hz, CH), 117.67 (d, *J*$_{C\text{-}F}$ = 19.6 Hz, CH), 73.74 (d, *J*$_{C\text{-}F}$ = 2.6 Hz, C), 45.29 (d, *J*$_{C\text{-}F}$ = 4.7 Hz, CH$_2$), 25.22 (CH$_3$); HRMS (ESI) *m/z*: 272.1077 [M + H]$^+$, C$_{16}$H$_{15}$FNO$_2$ required 272.1070.

*(S)-1-Benzyl-7-chloro-3-hydroxy-3-methylindolin-2-one* (**2p**): Enantiomeric ratio (83:17) was determined by chiral HPLC (Chiralpak AD-H), hexane-iPrOH 80:20, 1.0 mL/min, major enantiomer rt = 9.0 min, minor enantiomer rt = 7.3 min. White solid; mp = 175–176 °C; $[\alpha]_{20}^{D}$ = −18.92 (c = 0.945, CHCl$_3$) (83:17 er); $^1$H NMR (300 MHz, CDCl3) δ 7.40–7.15 (m, 7H), 7.02 (dd, *J* = 8.2, 7.3 Hz, 1H), 5.32 (s, 2H), 3.25 (s, 1H), 1.66 (s, 3H). $^{13}$C NMR (75 MHz, CDCl$_3$) δ 179.35 (C), 137.95 (C), 137.07 (C), 134.31 (C), 132.02 (CH), 128.61 (CH), 127.24 (CH), 126.33 (CH), 124.30 (CH), 122.15 (CH), 115.87 (C), 73.06 (C), 44.75 (CH$_2$), 25.42 (CH$_3$); HRMS (ESI) *m/z*: 288.0783 [M + H]$^+$, C$_{16}$H$_{15}$ClNO$_2$ required 288.0786.

*(S)-1-Benzyl-3-hydroxy-3,5,7-trimethylindolin-2-one* (**2q**): Enantiomeric ratio (89:11) was determined by chiral HPLC (Chiralpak AD-H), hexane-iPrOH 80:20, 1.0 mL/min, major enantiomer rt = 9.6 min, minor enantiomer rt = 7.6 min. White solid; mp = 142–145 °C; $[\alpha]_{20}^{D}$ = −37.19 (c = 0.855, CHCl$_3$) (89:11 er); $^1$H NMR (300 MHz, CDCl$_3$) δ 7.38–7.20 (m, 3H), 7.16–7.11 (m, 3H), 6.78 (dq, *J* = 1.7, 0.7 Hz, 1H), 5.17 (d, *J* = 16.6 Hz, 1H), 5.10 (d, *J* = 16.6 Hz, 1H), 3.14 (s, 1H), 2.28 (s, 3H), 2.20 (s, 3H), 1.67 (s, 3H). $^{13}$C RMN (75 MHz, CDCl$_3$) δ 179.72 (C), 137.33 (C), 137.25 (C), 133.80 (CH), 133.00 (C), 132.22 (C), 128.85 (CH), 127.20 (CH), 125.57 (CH), 122.15 (CH), 120.05 (C), 73.03 (C), 44.84 (CH$_2$), 25.49 (CH$_3$), 20.66 (CH$_3$), 18.50 (CH$_3$); HRMS (ESI) *m/z*: 282.1485 [M + H]$^+$, C$_{18}$H$_{20}$NO$_2$ required 282.1489.

*3.3. Procedures and Characterization Data for Compounds* **3a** *and* **4a**

*(1aS,6bS)-2-benzyl-6b-methyl-1a,6b-dihydro-2H-oxireno[2,3-b]indole* (**3a**): A 1 M LiAlH$_4$ solution in THF (0.2 mL, 0.2 mmol) was added dropwise on a solution of **2a** (0.1 mmol) in THF (5 mL) at room temperature under nitrogen. The reaction was warmed to 75 °C and stirred until the reaction was complete (TLC). The reaction mixture was quenched with NH$_4$Cl (10 mL), extracted with dichloromethane (3 × 20 mL), washed with brine (10 mL), dried over MgSO$_4$ and dried under reduced pressure. The crude was used for the next step without further purification. $^1$H NMR (300 MHz, CDCl$_3$) δ 7.36–7.14 (m, 6H), 7.05 (td, *J* = 7.7, 1.3 Hz, 1H), 6.67 (ddt, *J* = 8.2, 7.4, 0.8 Hz, 1H), 6.34 (dd, *J* = 7.8, 0.8 Hz, 1H), 4.54 (s, 1H), 4.45 (d, *J* = 15.6 Hz, 1H), 4.23 (d, *J* = 15.7 Hz, 1H), 1.47 (s, 3H). $^{13}$C NMR (75 MHz, CDCl$_3$) δ 148.45 (C), 138.10 (C), 131.72 (C), 129.94 (CH), 128.80 (CH), 127.17 (CH), 127.00 (CH), 123.15 (CH), 118.79 (CH), 107.51 (CH), 92.77 (CH), 75.79 (C), 48.51 (CH$_2$), 24.33 (CH$_3$).

*(2R,3S)-1-benzyl-3-hydroxy-3-methylindoline-2-carbonitrile* (**4a**): TMSCN (37 μL, 0.294 mmol) was added dropwise on a solution of **3a** (0.1 mmol) in CH$_2$Cl$_2$ (2 mL) at room temperature under nitrogen.

The reaction was stirred until the reaction was complete (TLC). Finally, the reaction mixture was directly poured into the column chromatography, using hexanes:EtOAc (95:5) as eluent to afford product **4a**. Enantiomeric ratio (89:11) was determined by chiral HPLC (Chiralpak AD-H), hexane-iPrOH 80:20, 1.0 mL/min, major enantiomer rt = 7.9 min, minor enantiomer rt = 18.7 min. Oil; $[\alpha]_{20}^{D}$ = −46.57 (c = 0.505, CHCl$_3$) (89:11 er); $^1$H NMR (300 MHz, CDCl$_3$) δ 7.43–7.19 (m, 7H), 6.89 (td, J = 7.5, 0.9 Hz, 1H), 6.68 (dt, J = 8.1, 0.7 Hz, 1H), 4.71 (d, J = 14.8 Hz, 1H), 4.19 (d, J = 14.9 Hz, 1H), 4.04 (s, 1H), 2.55 (s, 1H), 1.64 (s, 3H); $^{13}$C NMR (75 MHz, CDCl$_3$) δ 148.38 (C), 135.77 (C), 132.15 (C), 130.38 (CH), 128.90 (CH), 128.25 (CH), 128.01 (CH), 122.89 (CH), 120.36 (CH), 115.48 (C), 109.24 (CH), 78.41 (C), 66.88 (CH), 50.95 (CH$_2$), 25.36 (CH$_3$); HRMS (ESI) m/z: 265.1329 [M + H]$^+$, C$_{17}$H$_{17}$N$_2$O required 265.1335.

## 4. Conclusions

We have developed a catalytic enantioselective addition of Me$_2$Zn to isatins catalyzed by a chiral Zn(II) complex using as chiral ligand a α-hydroxyamide derived from (S)-mandelic acid. The corresponding chiral 3-hydroxy-3-methyl-2-oxindoles are obtained with good yields and enantioselectivities. The enantioselectivities are comparable to the example described by Shibashaki [34] with a bifunctional proline-derived amino alcohol. The advantages of our system are that the catalyst is easily prepared in a one-step procedure, the reaction time is shorter and no slow addition of the reagent is required, leading to simplified procedures. Moreover, several transformations have been done with the corresponding chiral tertiary alcohols obtained.

**Supplementary Materials:** The following are available online at www.mdpi.com/2073-4344/7/12/387/s1, $^1$H and $^{13}$C NMR spectra, and HPLC chromatograms of all compounds.

**Acknowledgments:** Financial support from the MINECO (Ministerio de Economía, Industria y Competitividad, Gobierno de España; CTQ2013-47494-P). C.V. thanks MINECO for a JdC contract. Access to NMR and MS (Mass Spectrometry) facilities from the Servei central de suport a la investigació experimental (SCSIE)-UV is also acknowledged.

**Author Contributions:** C.V. and J.R.P. conceived and designed the experiments; A.d.C. performed the experiments; C.V. and A.d.C. analyzed the data; G.B. contributed reagents/materials/analysis tools; C.V. and J.R.P wrote the paper. All authors read, revised and approved the final manuscript.

**Conflicts of Interest:** The authors declare no conflict of interest.

## References and Note

1. Labroo, R.B.; Cohen, L.A. Preparative separation of the diastereoisomers of dioxindolyl-L-alanine and assignment of stereochemistry at C-3. *J. Org. Chem.* **1990**, *55*, 4901–4904. [CrossRef]
2. Koguchi, Y.; Kohno, J.; Nishio, M.; Takahashi, K.; Okuda, T.; Ohnuki, T.; Komatsubara, S. TMC-95A, B, C, and D, Novel Proteasome Inhibitors Produced by *Apiospora montagnei* Sacc. TC 1093 Taxonomy, Production, Isolation, and Biological Activities. *J. Antibiot.* **2000**, *53*, 105–109. [CrossRef] [PubMed]
3. Tang, Y.-Q.; Sattler, I.; Thiericke, R.; Grabley, S.; Feng, X.-Z. Maremycins C and D, New Diketopiperazines, and Maremycins E and F, Novel Polycyclic spiro-Indole Metabolites Isolated from *Streptomyces* sp. *Eur. J. Org. Chem.* **2001**, *2001*, 261–267. [CrossRef]
4. Hibino, S.; Choshi, T. Simple indole alkaloids and those with a nonrearranged monoterpenoid unit. *Nat. Prod. Rep.* **2001**, *18*, 66–87. [CrossRef] [PubMed]
5. Albrecht, B.K.; Williams, R.M. A concise, total synthesis of the TMC-95A/B proteasome inhibitors. *Proc. Natl. Acad. Sci. USA* **2004**, *101*, 11949–11954. [CrossRef] [PubMed]
6. Peddibhotla, S. 3-Substituted-3-hydroxy-2-oxindole, an Emerging New Scaffold for Drug Discovery with Potential Anti-Cancer and other Biological Activities. *Curr. Bioact. Compd.* **2009**, *5*, 20–38. [CrossRef]
7. Erkizan, H.V.; Kong, Y.; Merchant, M.; Schlottmann, S.; Barber-Rotenberg, J.S.; Yuan, L.; Abaan, O.D.; Chou, T.-H.; Dakshanamurthy, S.; Brown, M.L.; et al. A small molecule blocking oncogenic protein EWS-FLI1 interaction with RNA helicase A inhibits growth of Ewing's sarcoma. *Nat. Med.* **2009**, *15*, 750–756. [CrossRef] [PubMed]

8. Hewawasam, P.; Erway, M.; Moon, S.L.; Knipe, J.; Weiner, H.; Boissard, C.G.; Post-Munson, D.J.; Gao, Q.; Huang, S.; Gribkoff, V.K.; et al. Synthesis and Structure–Activity Relationships of 3-Aryloxindoles: A New Class of Calcium-Dependent, Large Conductance Potassium (Maxi-K) Channel Openers with Neuroprotective Properties. *J. Med. Chem.* **2002**, *45*, 1487–1499. [CrossRef] [PubMed]

9. Tokunaga, T.; Hume, W.E.; Umezome, T.; Okazaki, K.; Ueki, Y.; Kumagai, K.; Hourai, S.; Nagamine, J.; Seki, H.; Taiji, M.; et al. Oxindole Derivatives as Orally Active Potent Growth Hormone Secretagogues. *J. Med. Chem.* **2001**, *44*, 4641–4649. [CrossRef] [PubMed]

10. Boechat, N.; Kover, W.B.; Bongertz, V.; Bastos, M.M.; Romeiro, N.C.; Azavedo, M.L.G.; Wollinger, W. Design, Synthesis and Pharmacological Evaluation of HIV-1 Reverse Transcriptase Inhibition of New Indolin-2-Ones. *Med. Chem.* **2007**, *3*, 533–542. [CrossRef] [PubMed]

11. Barber-Rotenberg, J.S.; Selvanathan, S.P.; Kong, Y.; Erkizan, H.V.; Snyder, T.M.; Hong, S.P.; Kobs, C.L.; South, N.L.; Summer, S.; Monroe, P.J.; et al. Single Enantiomer of YK-4-279 Demonstrates Specificity in Targeting the Oncogene EWS-FLI1. *Oncotarget* **2012**, *3*, 172–182. [CrossRef] [PubMed]

12. Kumar, A.; Chimni, S.S. Catalytic asymmetric synthesis of 3-hydroxyoxindole: A potentially bioactive molecule. *RSC Adv.* **2012**, *2*, 9748–9762. [CrossRef]

13. Yu, B.; Xing, H.; Yu, D.-Q.; Liu, H.-M. Catalytic asymmetric synthesis of biologically important 3-hydroxyoxindoles: An update. *Beilstein J. Org. Chem.* **2016**, *12*, 1000–1039. [CrossRef] [PubMed]

14. Qiao, X.-C.; Zhu, S.-F.; Zhou, Q.-L. From allylic alcohols to chiral tertiary homoallylic alcohol: Palladium-catalyzed asymmetric allylation of isatins. *Tetrahedron Asymmetry* **2009**, *20*, 1254–1261. [CrossRef]

15. Hanhan, N.V.; Tang, Y.C.; Tran, N.T.; Franz, A.K. Scandium(III)-Catalyzed Enantioselective Allylation of Isatins Using Allylsilanes. *Org. Lett.* **2012**, *14*, 2218–2221. [CrossRef] [PubMed]

16. Itoh, J.; Han, S.; Krische, M.J. Enantioselective Allylation, Crotylation, and Reverse Prenylation of Substituted Isatins: Iridium-Catalyzed C-C Bond-Forming Transfer Hydrogenation. *Angew. Chem. Int. Ed.* **2009**, *48*, 6313–6316. [CrossRef] [PubMed]

17. Shintani, R.; Inoue, M.; Hayashi, T. Rhodium-Catalyzed Asymmetric Addition of Aryl- and Alkenylboronic Acids to Isatins. *Angew. Chem. Int. Ed.* **2006**, *45*, 3353–3356. [CrossRef] [PubMed]

18. Toullec, P.Y.; Jagt, R.B.C.; de Vries, J.G.; Feringa, B.L.; Minnaard, A.J. Rhodium-Catalyzed Addition of Arylboronic Acids to Isatins: An Entry to Diversity in 3-Aryl-3-Hydroxyoxindoles. *Org. Lett.* **2006**, *8*, 2715–2718. [CrossRef] [PubMed]

19. Gajulapalli, V.P.R.; Vinayagam, P.; Kesavan, V. Organocatalytic asymmetric decarboxylative cyanomethylation of isatins using L-proline derived bifunctional thiourea. *Org. Biomol. Chem.* **2014**, *12*, 4186–4191. [CrossRef] [PubMed]

20. Yin, L.; Kanai, M.; Shibasaki, M. A Facile Pathway to Enantiomerically Enriched 3-Hydroxy-2-Oxindoles: Asymmetric Intramolecular Arylation of α-Keto Amides Catalyzed by a Palladium–DifluorPhos Complex. *Angew. Chem. Int. Ed.* **2011**, *50*, 7620–7623. [CrossRef] [PubMed]

21. Zhang, H.P.; Kamano, Y.; Ichihara, Y.; Kizu, H.; Komiyama, K.; Itokawa, H.; Pettit, G.R. Isolation and structure of convolutamydines B-D from marine bryozoan *Amathia convoluta*. *Tetrahedron* **1995**, *51*, 5523–5528. [CrossRef]

22. Totobenazara, J.; Bacalhau, P.; San Juna, A.A.; Marques, C.S.; Fernandes, L.; Goth, A.; Caldeira, A.T.; Martins, R.; Burke, A.J. Design, Synthesis and Bioassays of 3-Substituted-3-Hydroxyoxindoles for Cholinesterase Inhibition. *ChemistrySelect* **2016**, *1*, 3580–3588. [CrossRef]

23. Sumiyoshi, T.; Takahashi, Y.; Uruno, Y.; Takai, K.; Suwa, A.; Murata, Y. Novel Fused-Ring Pyrrolidine Derivative. Patent WO 2013122107 A1, 22 August 2012.

24. Currie, K.S.; Du, Z.; Farand, J.; Guerrero, J.A.; Katana, A.A.; Kato, D.; Lazerwith, S.E.; Li, J.; Link, J.O.; Mai, N.; et al. Syk Inhibitors. Patent WO 2015017610 A1, 5 Febrary 2015.

25. Ishimaru, T.; Shibata, N.; Nagai, J.; Nakamura, S.; Toru, T.; Kanemasa, S. Lewis Acid-Catalyzed Enantioselective Hydroxylation Reactions of Oxindoles and β-Keto Esters Using DBFOX Ligand. *J. Am. Chem. Soc.* **2006**, *128*, 16488–16489. [CrossRef] [PubMed]

26. Sano, D.; Nagata, K.; Itoh, T. Catalytic Asymmetric Hydroxylation of Oxindoles by Molecular Oxygen Using a Phase-Transfer Catalyst. *Org. Lett.* **2008**, *10*, 1593–1595. [CrossRef] [PubMed]

27. Yang, Y.; Moinodeen, F.; Chin, W.; Ma, T.; Jiang, Z.; Tan, C.-H. Pentanidium–Catalyzed Enantioselective α-Hydroxylation of Oxindoles Using Molecular Oxygen. *Org. Lett.* **2012**, *14*, 4762–4765. [CrossRef] [PubMed]

28. DiMauro, E.F.; Kozlowski, M.C. The First Catalytic Asymmetric Addition of Dialkylzincs to α-Ketoesters. *Org. Lett.* **2002**, *4*, 3781–3784. [CrossRef] [PubMed]

29. Wieland, L.C.; Deng, H.; Snapper, M.L.; Hoveyda, H. Al-Catalyzed Enantioselective Alkylation of α-Ketoesters by Dialkylzinc Reagents. Enhancement of Enantioselectivity and Reactivity by an Achiral Lewis Base Additive. *J. Am. Chem. Soc.* **2005**, *127*, 15453–15456. [CrossRef] [PubMed]

30. Fennie, M.W.; DiMauro, E.F.; O'Brien, E.M.; Annamalai, V.; Kozlowski, M.C. Mechanism and scope of salen bifunctional catalysts in asymmetric aldehyde and α-ketoester alkylation. *Tetrahedron* **2005**, *61*, 6249–6265. [CrossRef]

31. Wu, H.-L.; Wu, P.-Y.; Shen, Y.-Y.; Uang, B.-J. Asymmetric Addition of Dimethylzinc to α-Ketoesters Catalyzed by (−)-MITH. *J. Org. Chem.* **2008**, *73*, 6445–6447. [CrossRef] [PubMed]

32. Zheng, B.; Hou, S.; Li, Z.; Guo, H.; Zhong, J.; Wang, M. Enantioselective synthesis of quaternary stereogenic centers through catalytic asymmetric addition of dimethylzinc to α-ketoesters with chiral *cis*-cyclopropane-based amide alcohol as ligand. *Tetrahedron Asymmetry* **2009**, *20*, 2125–2129. [CrossRef]

33. Infante, R.; Nieto, J.; Andrés, C. Highly Homogeneous Stereocontrolled Construction of Quaternary Hydroxyesters by Addition of Dimethylzinc to α-Ketoesters Promoted by Chiral Perhydrobenzoxazines and B(OEt)$_3$. *Chem. Eur. J.* **2012**, *18*, 4375–4379. [CrossRef] [PubMed]

34. Funabashi, K.; Jachmann, M.; Kanai, M.; Shibasaki, M. Multicenter Strategy for the Development of Catalytic Enantioselective Nucleophilic Alkylation of Ketones: Me$_2$Zn Addition to α-Ketoesters. *Angew. Chem. Int. Ed.* **2003**, *42*, 5489–5492. [CrossRef] [PubMed]

35. Blay, G.; Fernández, I.; Hernández-Olmos, V.; Marco-Aleixandre, A.; Pedro, J.R. Enantioselective addition of dimethylzinc to aldehydes catalyzed by *N*-substituted mandelamide-Ti(IV) complexes. *Tetrahedron Asymmetry* **2005**, *16*, 1953–1958. [CrossRef]

36. Blay, G.; Fernández, I.; Marco-Aleixandre, A.; Pedro, J.R. Catalytic Asymmetric Addition of Dimethylzinc to α-Ketoesters, Using Mandelamides as Ligands. *Org. Lett.* **2006**, *8*, 1287–1290. [CrossRef] [PubMed]

37. Blay, G.; Fernández, I.; Marco-Aleixandre, A.; Pedro, J.R. Mandelamide−Zinc-Catalyzed Enantioselective Alkyne Addition to Heteroaromatic Aldehydes. *J. Org. Chem.* **2006**, *71*, 6674–6677. [CrossRef] [PubMed]

38. Blay, G.; Cardona, L.; Fernández, I.; Marco-Aleixandre, A.; Muñoz, M.C.; Pedro, J.R. Catalytic enantioselective addition of terminal alkynes to aromatic aldehydes using zinc-hydroxyamide complexes. *Org. Biomol. Chem.* **2009**, *7*, 4301–4308. [CrossRef] [PubMed]

39. Blay, G.; Fernández, I.; Hernández-Olmos, V.; Marco-Aleixandre, A.; Pedro, J.R. Tailoring the ligand structure to the reagent in the mandelamide-Ti(IV) catalyzed enantioselective addition of dimethyl- and diethylzinc to aldehydes. *J. Mol. Catal. A Chem.* **2007**, *276*, 235–243. [CrossRef]

40. Blay, G.; Fernández, I.; Marco-Aleixandre, A.; Pedro, J.R. Enantioselective Addition of Dimethylzinc to α-Keto Esters. *Synthesis* **2007**, 3754–3757. [CrossRef]

41. The establishment of the absolute configuration of compound **2c** was determined by chemical correlation (reference 34).

42. Yan, W.; Wang, D.; Feng, J.; Li, P.; Zhao, D.; Wang, R. Synthesis of *N*-Alkoxycarbonyl Ketimines Derived from Isatins and Their Application in Enantioselective Synthesis of 3-Aminooxindoles. *Org. Lett.* **2012**, *14*, 2512–2515. [CrossRef] [PubMed]

43. Liu, M.; Zhang, C.; Ding, M.; Tang, B.; Zhang, F. Metal-free base-mediated oxidative annulation cascades to 3-substituted-3-hydroxyoxindole and its 3-spirocyclic derivative. *Green Chem.* **2017**, *19*, 4509–4514. [CrossRef]

44. Li, D.; Yu, W. Oxygen-Involved Oxidative Deacetylation of α-Substituted β-Acetyl Amides—Synthesis of α-Keto Amides. *Adv. Synth. Catal.* **2013**, *355*, 3708–3714. [CrossRef]

45. Lu, S.; Poh, S.B.; Siau, W.-Y.; Zhao, Y. Kinetic Resolution of Tertiary Alcohols: Highly Enantioselective Access to 3-Hydroxy-3-Substituted Oxindoles. *Angew. Chem. Int. Ed.* **2013**, *52*, 1731–1734. [CrossRef] [PubMed]

46. England, D.B.; Merey, G.; Padwa, A. Substitution and Cyclization Reactions Involving the Quasi-Antiaromatic 2H-Indol-2-one Ring System. *Org. Lett.* **2007**, *9*, 3805–3807. [CrossRef] [PubMed]

47. Monde, K.; Taniguchi, T.; Miura, N.; Nishimura, S.-I.; Harada, N.; Dukor, R.K.; Nafie, L.A. Preparation of cruciferous phytoalexin related metabolites, (−)-dioxibrassinin and (−)-3-cyanomethyl-3-hydroxyoxindole, and determination of their absolute configurations by vibrational circular dichroism (VCD). *Tetrahedron Lett.* **2003**, *44*, 6017–6020. [CrossRef]

48. Chen, Y.-S.; Cheng, M.-J.; Hsiao, Y.; Chan, H.-Y.; Hsieh, S.-Y.; Chang, C.-W.; Liu, T.-W.; Chang, H.-S.; Chen, I.-S. Chemical Constituents of the Endophytic Fungus *Hypoxylon* sp. 12F0687 Isolated from Taiwanese *Ilex formosana*. *Helv. Chim. Acta* **2015**, *98*, 1167–1176. [CrossRef]

49. Gorokhovik, I.; Neuville, L.; Zhu, J. Trifluoroacetic Acid-Promoted Synthesis of 3-Hydroxy, 3-Amino and Spirooxindoles from α-Keto-*N*-Anilides. *Org. Lett.* **2011**, *13*, 5536–5539. [CrossRef] [PubMed]
50. Sarraf, D.; Richy, N.; Vidal, J. Synthesis of Lactams by Isomerization of Oxindoles Substituted at C-3 by an ω-Amino Chain. *J. Org. Chem.* **2014**, *79*, 10945–10955. [CrossRef] [PubMed]

*catalysts*

MDPI

*Article*

# Biotransformation of Ergostane Triterpenoid Antcin K from *Antrodia cinnamomea* by Soil-Isolated *Psychrobacillus* sp. AK 1817

Chien-Min Chiang [1], Tzi-Yuan Wang [2], An-Ni Ke [3], Te-Sheng Chang [3,*] and Jiumn-Yih Wu [4,*]

[1] Department of Biotechnology, Chia Nan University of Pharmacy and Science, No. 60, Sec. 1, Erh-Jen Rd., Jen-Te District, Tainan 71710, Taiwan; cmchiang@mail.cnu.edu.tw

[2] Biodiversity Research Center, Academia Sinica, Taipei 115, Taiwan; tziyuan@gmail.com

[3] Department of Biological Sciences and Technology, National University of Tainan, No. 33, Sec. 2, Shu-Lin St., Tainan 70005, Taiwan; koannie84617@gmail.com

[4] Department of Food Science, National Quemoy University, No. 1, University Road., Jin-Ning Township, Kinmen County 892, Taiwan

* Correspondence: mozyme2001@gmail.com (T.-S.C.); wujy@nqu.edu.tw (J.-Y.W.); Tel.: +886-6-2606153 (T.-S.C.); +886-82-313310 (J.-Y.W.); Fax: +886-6-2606153 (T.-S.C.); +886-82-313797 (J.-Y.W.)

Received: 22 September 2017; Accepted: 3 October 2017; Published: 11 October 2017

**Abstract:** Antcin K is one of the major ergostane triterpenoids from the fruiting bodies of *Antrodia cinnamomea*, a parasitic fungus that grows only on the inner heartwood wall of the aromatic tree *Cinnamomum kanehirai* Hay (Lauraceae). To search for strains that have the ability to biotransform antcin K, a total of 4311 strains of soil bacteria were isolated, and their abilities to catalyze antcin K were determined by ultra-performance liquid chromatography analysis. One positive strain, AK 1817, was selected for functional studies. The strain was identified as *Psychrobacillus* sp., based on the DNA sequences of the 16S rRNA gene. The biotransformation metabolites were purified with the preparative high-performance liquid chromatography method and identified as antcamphin E and antcamphin F, respectively, based on the mass and nuclear magnetic resonance spectral data. The present study is the first to report the biotransformation of triterpenoids from *A. cinnamomea* (*Antrodia cinnamomea*).

**Keywords:** *Antrodia cinnamomea*; biotransformation; antcin K; triterpenoid; *Psychrobacillus*

---

## 1. Introduction

Terpenoids, which are composed of isoprenoid subunits, are the largest group of phytochemicals in the world, and are widely distributed in almost all living organisms [1]. Based on the number of isoprenoid units, terpenoids are subdivided into monoterpenoids (C10), sesquiterpenoids (C15), diterpenoids (C20), sesterterpenoids (C25), triterpenoids (C30), and tetraterpenoids (C40). Triterpenoids from several sources are used for medicinal purposes in Asia as anti-inflammatory, analgesic, antipyretic, hepatoprotective, and cardiotonic agents.

*Antrodia cinnamomea*, also called *A. camphorate*, is a parasitic fungus that grows only on the inner heartwood wall of the aromatic tree *Cinnamomum kanehirai* Hay (Lauraceae). This mushroom is endemic in Taiwan, and is used as a folk remedy in the treatment of a variety of diseases [2]. The fruiting bodies of *A. cinnamomea* contain abundant ergostane and lanostane tetracyclic triterpenoids, which are generally considered the major bioactive constituents of *A. cinnamomea* (*Antrodia cinnamomea*). More than 40 triterpenoids have been isolated from *A. cinnamomea*, where antcin K is one of the major ergostane triterpenoids from the fruiting bodies of cutting wood-cultivated, as well as dish-cultivated, *A. cinnamomea*. Some bioactivities of antcin K have been reported, including anti-inflammatory, antidiabetic, and antihyperlipidemic activities, inducing apoptosis of hepatoma cells, and reducing carcinogenesis [3–8].

Biotransformation of xenobiotics using microorganisms is a very useful approach for expanding the chemical diversity of natural products [9]. Many biotransformations of triterpenoids, including boswellic acid [10,11], oleanolic acid [12,13], betulinic acid [12,14], maslinic acid [13,15], ursolic acid [16–18], and ginseng-containing triterpenoids, such as ginsenosides [19], protopanaxadiol [20,21], and dipterocarpol [22], have been discovered. The biocatalysts used in studies include the fungi *Aspergillus, Cunninghamella, Absidia, Rhizomucor,* and *Syncephalastrum* and the bacteria *Bacillus* and *Nocardia*. The biotransformation reactions include hydroxylation, dehydrogenation, lactone formation, methylation, and (de)glycosylation. Many novel triterpenoids with potent bioactivity have been identified from the biotransformation reactions; however, to our knowledge, biotransformation of triterpenoids from *A. cinnamomea* has not been realized yet, although many bioactivities of triterpenoids have been identified from *A. cinnamomea*. Based on the abundance of antcin K in dish-cultured *A. cinnamomea*, biotransformation of antcin K is a good starting point with an alternate high-throughput screening method. In the present study, thousands of soil bacteria were isolated, and the ability to catalyze antcin K was determined. One positive strain was selected and then identified by genetic analysis. The biotransformation metabolites were purified with preparative high-performance liquid chromatography (HPLC), and identified using spectra methods.

## 2. Results

### 2.1. Screening and Identification of Soil Bacteria with Biotransformation Activity

A great challenge for the realization of a desired biotransformation reaction is finding the appropriate microorganism. Thus, classical screening of a series of microbial strains is still the most widely used technique. To study the biotransformation of triterpenoid antcin K from *A. cinnamomea*, thousands of soil bacteria were isolated with the plating method and then cultivated in broth with antcin K. The fermentation broth was analyzed using ultra-performance liquid chromatography (UPLC) to determine the ability of the strain to digest antcin K. A total of 4311 strains were screened, and one strain (AK 1817) was selected for functional studies.

Figure 1a shows the UPLC analysis of 0 h and 72 h fermentation broths of the strain. In the figure, the precursors 25*S*-antcin K and 25*R*-antcin K appear with retention times of 4.6 and 4.7 min, respectively, for the 0 h fermentation broth. After 72 h cultivation, 25*S*-antcin K and 25*R*-antcin K decreased dramatically, and two new major peaks, compound (**1**) and compound (**2**), appeared with retention times of 5.3 and 5.4 min, respectively. Parallel experiments were repeated with the strain AK 1817, but without antcin K in the fermentation broth. No metabolites with a retention time of 5.3 and 5.4 min appeared in the 72 h fermentation broth (Figure S1). From the results, it is clear that antcin K was catalyzed by the strain to compound (**1**) and compound (**2**). Moreover, compound (**1**) and compound (**2**) were potentially biotransformed from 25*S*-antcin K and 25*R*-antcin K, respectively.

To evaluate the biotransformation process in advance, we isolated 25*S*-antcin K and 25*R*-antcin K, and repeated the biotransformation experiments with the strain AK 1817 by adding 25*S*-antcin K or 25*R*-antcin K individually in the fermentation broth. Figure 1b,c shows the 0 h and 72 h fermentation broths of the strain fed with 25*S*-antcin K and 25*R*-antcin K, respectively. From the results, it is clear that compound (**1**) was biotransformed from 25*S*-antcin K (Figure 1b), and compound (**2**) was biotransformed from 25*R*-antcin K (Figure 1c). According to the absorption in the UPLC analysis, the conversions of compound (**1**) and compound (**2**) from 25*S*-antcin K and 25*R*-antcin K at the 72-h cultivation were 67.6% and 74.8%, respectively.

To identify the strain, the partial 16S rRNA gene was amplified and sequenced with polymerase chain reaction (PCR) with the bacteria-specific 27F (5′-AGAGTTTGATCCTGGCTCAG-3′) and 1391R (5′-GACGGGCRGTGWGTRCA-3′) primer set. The DNA sequence is shown in Figure S2. The partial sequences of the 16S rRNA gene were then blasted against National Center for Biotechnology

Information (NCBI) non-redundant nucleotides to identify the strain. From the blasted results, the phylogenetic tree indicated that the AK 1817 strain was classified as *Psychrobacillus* sp. (Figure 2). The strain was deposited in the Bioresources Collection and Research Center (BCRC, Food Industry Research and Development Institute, Hsinchu, Taiwan).

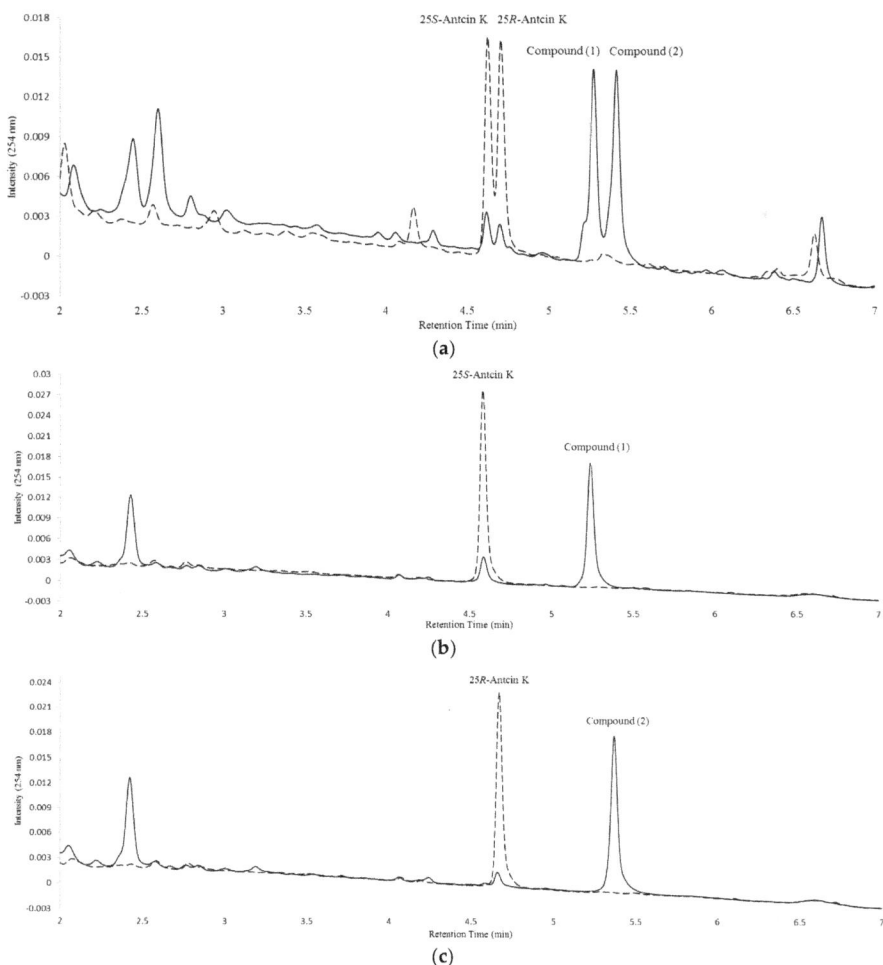

**Figure 1.** Biotransformation of antcin K by soil bacteria AK 1817 strain. The strain was cultivated in Luria-Bertani (LB) media containing (**a**) 25S- and 25R-antcin K; (**b**) 25S-antcin K; or (**c**) 25R-antcin K. The 0 h (dash curves) and 72 h (solid curves) cultivation of the fermentation broth were analyzed with UPLC. The UPLC operation conditions are described in Materials and Methods.

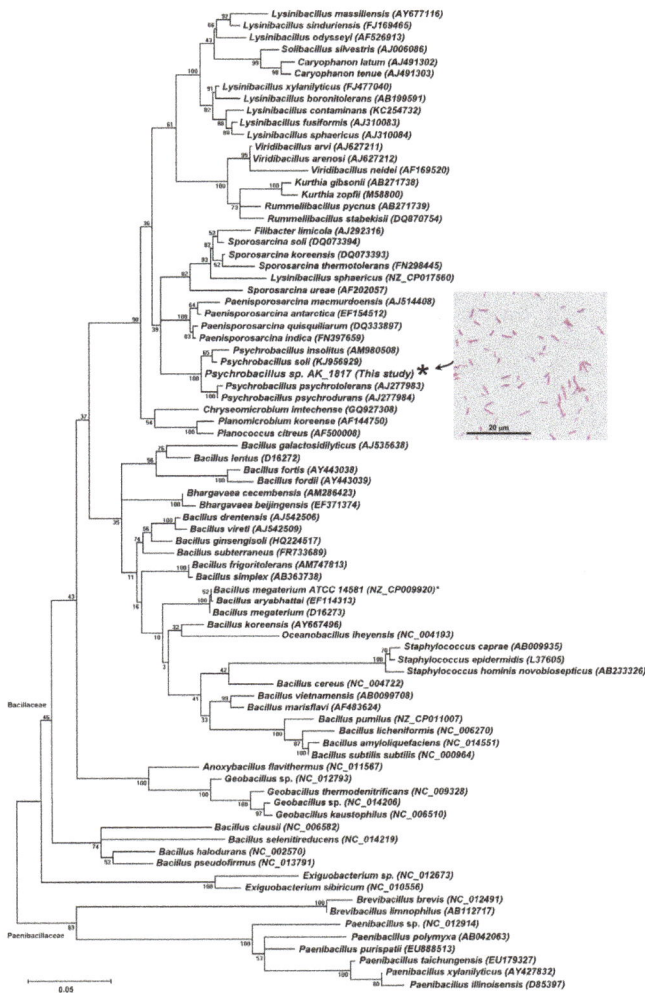

**Figure 2.** Molecular phylogenetic analysis of strain AK 1817 with the maximum likelihood (ML) method based on the general time reversible (GTR) model [23]. The tree with the highest log likelihood (−11,124.2673) is shown. The percentage of trees in which the associated taxa clustered together is shown next to the branches. The initial tree(s) for the heuristic search were obtained automatically by applying neighbor-join and BioNJ algorithms to a matrix of pairwise distances, estimated using the maximum composite likelihood (MCL) approach, and then selecting the topology with the superior log likelihood value. A discrete Gamma distribution was used to model evolutionary rate differences among sites (six categories (+G, parameter = 0.1906)). The rate variation model allowed for some sites to be evolutionarily invariable ([+I], 60.3946% sites). The tree is drawn to scale, with branch lengths measured in the number of substitutions per site. The analysis involved 82 nucleotide sequences. All positions containing gaps and missing data were eliminated. There were a total of 1146 positions in the final dataset. Evolutionary analyses were conducted in Molecular Evolutionary Genetics Analysis (MEGA) version 7.0 [24]. Gram staining revealed strain AK 1817 is a short-rod-shaped, Gram-positive bacteria (see the inserted photo).

*2.2. Isolation and Identification of Biotransformation Metabolites*

To isolate the two metabolites, the fermentation was scaled up with a 5 L fermenter. Two batches of 2.5 L fermentation were performed in the 5 L fermenter. Compound (**1**) and compound (**2**) were further isolated using the preparative HPLC method, and were identified using spectrophotometric methods. Compound (**1**) and compound (**2**) showed identical mass and nuclear magnetic resonance (NMR) spectrum data. They showed an $[M + H]^+$ ion peak at $m/z$: 487.42 in the electrospray ionization mass (ESI-MS) spectrum corresponding to the molecular formula $C_{29}H_{42}O_6$. Then $^1H$ and $^{13}C$ NMR, including distortionless enhancement by polarization transfer (DEPT), heteronuclear single quantum coherence (HSQC), heteronuclear multiple bond connectivity (HMBC), nuclear Overhauser effect spectroscopy (NOESY), and correlation spectroscopy (COSY), spectra were obtained, and the $^1H$- and $^{13}C$-NMR signal assignments were conducted accordingly (shown in Figures S3–S6). The key HMBC correlations of compounds (**1**) and (**2**) are shown in Figure S7, and the spectroscopic data is listed in Table S1. Based on these spectral data and with the comparison of $^1H$-NMR and $^{13}C$-NMR data in the literature [25], as well as the biotransformation results above, compound (**1**) and compound (**2**) were characterized as antcamphin E and antcamphin F, respectively. Figure 3 illustrates the biotransformation of antcin K by the soil-isolated *Psychrobacillus* sp. AK 1817 strain.

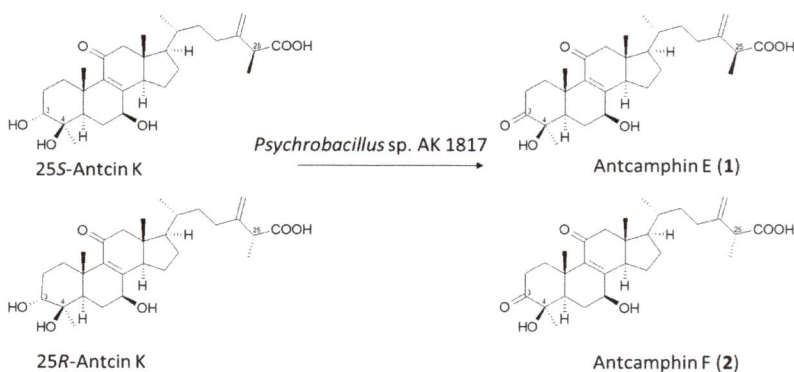

**Figure 3.** Biotransformation process of antcin K by the *Psychrobacillus* sp. AK 1817 strain.

## 3. Discussion

The present study demonstrated the biotransformation of 25*S*-antcin K and 25*R*-antcin K to antcamphin E and antcamphin F, respectively, by soil-isolated *Psychrobacillus* sp. AK 1817. To the best of our knowledge, this study is the first to report a microbe having direct biotransformation ability toward triterpenoids from *A. cinnamomea*.

In addition to the triterpenoids from *A. cinnamomea*, many biotransformations of triterpenoids from other sources have been reported [11–22,26–29]. In the present study, *Psychrobacillus* sp. AK 1817 catalyzed dehydrogenation (oxidation) of the C-3 hydroxyl group on the antcin K structure. Some bacteria have been reported to catalyze oxidation of a 3-hydroxyl group on the triterpenoid structure to form corresponding 3-oxo-triterpenoid derivatives. Among these bacteria, three Actinobacteria (*Mycobacterium* sp. [27], *Nocardia coralline* [28], and *Rhodococcus rhodochrous* [29]) have been demonstrated to catalyze the oxidation of the 3-hydroxyl group on the tetracyclic triterpenoid 20*R*-dihydroprotopanaxadiol and the pentacyclic triterpenoids echinocystic acid and betulin, respectively, to form the corresponding 3-oxo-triterpenoid derivatives. In addition, the *Bacillus megaterium* ATCC (American Type Culture Collection) 14581 and CGMCC (China General Microbiological Culture Collection Center) 1.1741 strains were found to catalyze the oxidation of the 3-hydroxyl group on the pentacyclic triterpenoids betulinic acid [26] and ursolic acid [18], respectively, to form the corresponding 3-oxo-triterpenoid derivatives. *B. megaterium* (*Bacillus megaterium*) ATCC 14581 is within the ancestral *Bacillus* clade while *Psychrobacillus* sp. AK 1817 is within the derived clade (Figure 2). Therefore, due to the similar biotransformation modifications by the two microbes, for comparison, we used the strain *B. megaterium* ATCC 14581 to perform the biotransformation experiments for antcin K. However, no metabolite was found in the reaction (Figure S7). The results revealed the unique substrate specificity in the biotransformation of antcin K by the strain *Psychrobacillus* sp. AK 1817. To resolve the corresponding catalyzing enzyme from *B. megaterium* ATCC 14581 and *Psychrobacillus* sp. AK 1817, a comparative functional genomic project with the two strains is being conducted in our laboratory.

Due to the abundance of antcin K in wood-cultivated, as well as in dish-cultivated, *A. cinnamomea*, many bioactivities of antcin K have been found [3–8]. In contrast, few studies on antcamphin E and antcamphin F have been reported, due to their rarity. Recently, antcamphin E and antcamphin F were isolated as minor components of *A. cinnamomea* [25,30]. Antcamphin E and antcamphin F exhibited cytotoxic activity toward breast cancer cells and lung cancer cells. In addition, both compounds exhibited protective activities against CCl$_4$-induced injury in HepG2 cells [31]. These results highlight the valuable application of the biotransformation process found in the present study. The biotransformation process of *Psychrobacillus* sp. AK 1817 provides an easy and unique approach for obtaining antcamphin E and antcamphin F from the abundant antcin K. Therefore, deeper and broader studies on the bioactivity of the two compounds should be conducted in the future. In addition, the process could be scaled up for industrial applications. Further studies to clone candidate genes that encode the corresponding catalyzing enzyme from *Psychrobacillus* sp. AK 1817, and to extend the substrate specificity, are being performed in our laboratory.

In conclusion, in the present study, triterpenoids were newly biotransformed from *A. cinnamomea*, and a biotransformation process was developed to produce antcamphin E and antcamphin F from antcin K.

## 4. Materials and Methods

### 4.1. Microorganism and Chemicals

*Bacillus megaterium* ATCC 14581 (BCRC 10608) was purchased from BCRC. Dimethyl sulfoxide (DMSO) was purchased from Sigma-Aldrich (St. Louis, MO, USA). All the materials needed for polymerase chain reaction (PCR), including primers, deoxyribonucleotide triphosphate, and Taq DNA polymerase, were purchased from MDBio (Taipei, Taiwan). The other reagents and solvents used were of high quality, and were purchased from commercially available sources.

### 4.2. Preparation of Antcin K

The dried dish-cultivated *A. cinnamomea* sample was obtained from Honest & Humble Biotechnology Co., Ltd. (New Taipei City, Taiwan). First, the sample was extracted with 50% methanol to obtain the ergostane-enriched fraction. Then, this fraction was subjected to preparative HPLC to obtain antcin K. 1.2 g (1.2%) of antcin K were obtained from 100 g of the dried dish-cultivated *A. cinnamomea*. The 25S- and 25R-epimers of antcin K were further purified with semipreparative HPLC using the reverse-phase C-18 column. 13.5 and 23.7 mg of 25S- and 25R-epimers of antcin K were obtained from 70 mg of antcin K. The 25S- and 25R-epimers of antcin K were identified by comparison with standards provided by Professor Min Ye (Peking University, Beijing, China).

### 4.3. Screening and Identification of Soil Bacteria with Biotransformation Activity

A series of soil samples, collected from local areas of Tainan city in Taiwan, were used for isolation of the bacterial strains. The fresh soil samples were plated according to the dilution plating method on Luria-Bertani (LB) agar [32]. After cultivation at 28 °C for 24 h, the colonies that formed on the plates were transferred to a deep 48-well microplate containing 1 mL of LB medium. The microplate was incubated at 180 rpm and 28 °C for 1 day. About 0.1 mL of the primary culture was then transferred to another microplate containing 1 mL of LB medium and 100 mg/L of antcin K for the secondary cultivation. The remaining primary culture was stored at −20 °C until the biotransformation activity assay was completed. The secondary cultivation was carried out at 180 rpm and 30 °C for 3 days. Then, an equal volume of ethanol was added to each well of the secondary cultivation microplate and shaken vigorously for 30 min at 28 °C. The cell debris was removed by centrifugation at 4800 rpm for 30 min. The supernatant from the extracted broth was assayed with UPLC to measure the biotransformation activity. To confirm the activity of the biotransformation-positive strains, the tested strains were cultivated in a 250-mL baffled Erlenmeyer flask containing 20 mL LB medium and 100 mg/L antcin K at 180 rpm and 28 °C for 5 days. Samples were collected daily and analyzed with UPLC to confirm the biotransformation activity. Candidate strains were re-purified and then reanalyzed by repeating the biotransformation activity assay.

### 4.4. UPLC Analysis

The mixtures of the biotransformation reactions were analyzed with a UPLC system (Acquity UPLC H-Class, Waters, Milford, MA, USA). The system was equipped with an analytic C18 reversed-phase column (Acquity UPLC BEH C18, 1.7 μm, 2.1 i.d. × 100 mm, Waters, Milford, MA, USA). To analyze both antcin K and the biotransformation products, a gradient elution using water (A) containing 1% (*v/v*) acetic acid and methanol (B) with a linear gradient for 3 min with 50% to 80% B and for another 4 min with 80% to 100% B was conducted at a flow rate of 0.3 mL/min, an injection volume of 0.2 μL, and absorbance detection at 254 nm.

### 4.5. Candidate Strain Classification via 16S rRNA Gene Analysis

For determination of the 16S ribosomal RNA gene sequences of the biotransformation–positive strain, chromosomal DNA was isolated using a Geno Plus Genomic DNA Extraction Miniprep System (Viogene, Taipei, Taiwan) according to the manufacturer's instructions. The 16S rRNA gene was amplified using PCR with the forward (5′-AGAGTTTGATCCTGGCTCAG-3′) and reverse (5′- GACGGGCRGTGWGTRCA-3′) primers known to amplify the 16S rRNA gene from a broad range of taxonomically different bacterial isolates [33]. PCR was performed with a total volume of 100 μL, which contained PCR buffer, 1 μg genomic DNA, 0.2 mM (each) deoxyribonucleotide triphosphate, 50 pmol (each) forward and reverse primers, and 2.5 U of Taq DNA polymerase. Amplification was performed for 35 cycles in a DNA thermal cycler, the ABI Prism 377 DNA Sequencer/Genetic Analyzer (Perkin-Elmer, Boston, MA, USA), employing the thermal profile according to Hugenholtz and Goebel [33]. The sequence of the amplified DNA fragment was determined by the DNA Sequencing Center of National Cheng Kung University in Tainan (Taiwan). The 16S rRNA was then blasted against NCBI non-redundant nucleotides to identify the strain. The 16S rRNA phylogeny was constructed to classify the strain. Related sequences were downloaded and aligned using the ClustalX program [34], followed by manual modifications. The maximum-likelihood (ML) tree was reconstructed using GTR+G+I distances as implemented in the Molecular Evolutionary Genetics Analysis (MEGA) version 7.0 package [24] with 100 bootstrap replications [35]. The substitution model (parameter) used to calculate the GTR+G+I distances was selected using Modeltest 3.7 [36].

### 4.6. Scale-Up Fermentation, Isolation, and Identification of the Biotransformation Products

The AK 1817 strain was cultured in 100 mL of LB medium for 24 h as a seed culture, which was inoculated into a 5 L fermenter containing 2.5 L LB medium supplemented with 100 mg/L antcin K, followed by cultivation with aeration (0.5, *v/v/m*) and agitation (280 rpm) at 28 °C. A 10 mL cultured medium was sampled at several different time intervals and analyzed with UPLC. The purification process was the same as in our previous work [37], and is described briefly below. Two batches of 2.5 L fermentation were performed for the purification of the biotransformation products. Following fermentation, the broth was condensed to 200 mL under a vacuum, and extracted twice by ethyl acetate.

*Catalysts* **2017**, *7*, 299

The extracts were further condensed and the residue was then suspended in 200 mL of 50% methanol. The suspension was filtrated through a 0.2 μm nylon membrane. The filtrate was injected into a preparative YoungLin HPLC system (YL9100, YL Instrument, Gyeonggi-do, South Korea). The system was equipped with a preparative C18 reversed-phase column (Inertsil, 10 μm, 20.0 i.d. × 250 mm, ODS 3, GL Sciences, Eindhoven, The Netherlands). The operational conditions for the preparative HPLC analysis were the same as those in the UPLC analysis. The elution corresponding to the peaks of the metabolites in the UPLC analysis were collected, concentrated under vacuum, and then lyophilized. Finally, 27.9 mg of compound (**1**) and 33.3 mg of compound (**2**) were obtained, and the structures of the compounds were confirmed with NMR and mass spectral analysis. The mass analysis was performed on a Finnigan LCQ Duo mass spectrometer (ThermoQuest Corp., San Jose, CA, USA) with electrospray ionization (ESI). $^1$H- and $^{13}$C-NMR, DEPT, HSQC, HMBC, COSY, and NOESY spectra were recorded on a Bruker AV-700 NMR spectrometer (Bruker Corp., Billerica, MA, USA) at ambient temperature. Standard pulse sequences and parameters were used for the NMR experiments, and all chemical shifts were reported in parts per million (ppm, δ).

**Supplementary Materials:** The following are available online at www.mdpi.com/2073-4344/7/10/299/s1. Table S1. NMR spectroscopic data for compound (**1**)/(**2**) (in pyridine-d5; 700 MHz), Figure S1. UPLC analysis of fermentation broth of the AK 1817 strain. The strain was cultivated in LB media without antcin K. The fermentation broth with cultivation of 72-h was analyzed by UPLC. The UPLC operation conditions were described in Materials and Methods, Figure S2. The partial 16S rRNA gene sequence of the AK 1817 strain. The partial 16S rRNA gene was amplified and sequenced by PCR with the bacteria specific 27F (5'-AGAGTTTGATCCTGGCTCAG-3') and 1391R (5'-GACGGGCRGTGWGTRCA-3') primer set. The PCR operation conditions were described in Materials and Methods, Figure S3. The $^1$H-NMR (700 MHz, Pyridine-d5) spectrum of compound (**2**), Figure S4. The $^{13}$C-NMR (700 MHz, Pyridine-d5) spectrum of compound (**2**), Figure S5. The HMBC (700 MHz, Pyridine-d5) spectrum of compound (**2**), Figure S6. The HSQC (700 MHz, Pyridine-d5) spectrum of compound (**2**), Figure S7. The key HMBC correlations of compound (**1**)/(**2**), Figure S8. Biotransformation of antcin K by B. megaterium ATCC 14581 strain. The strain was cultivated in LB media containing both 25S- and 25R-antcin K. The fermentation broth with cultivation of 72-h was analyzed by UPLC. The UPLC operation conditions were described in Materials and Methods.

**Acknowledgments:** This research was financially supported by grants from the National Scientific Council of Taiwan (Project No. MOST 105-2221-E-024-018-).

**Author Contributions:** Te-Sheng Chang and Jiumn-Yih Wu conceived and designed the experiments and wrote the paper; Chien-Min Chiang purified the metabolites and resolved the chemical structures of compounds (**1**) and (**2**). Tzi-Yuan Wang identified the isolated bacteria. An-Ni Ke performed the experiments for the biotransformation and anti-proliferative activity assay of the tested compounds.

**Conflicts of Interest:** The authors declare no conflicts of interest.

# References

1. Sultana, N.; Saify, Z.S. Enzymatic biotransformation of terpenes as bioactive agents. *J. Enzym. Inhib. Med. Chem.* **2013**, *28*, 1113–1128. [CrossRef] [PubMed]
2. Lu, M.C.; El-Shazly, M.; Wu, T.Y.; Du, Y.C.; Chang, T.T.; Chen, C.F.; Hsu, Y.M.; Lai, C.P.; Chang, F.R.; Wu, Y.C. Recent research and development of *Antrodia cinnamomea*. *Pharmacol. Ther.* **2013**, *139*, 124–156. [CrossRef] [PubMed]
3. Shen, Y.C.; Wang, Y.H.; Chou, Y.C.; Chen, C.F.; Lin, L.X.; Chang, T.T.; Tien, J.H.; Chou, C.J. Evaluation of the anti-inflammatory activity of zhankuic acids isolated from the fruiting bodies of *Antrodia camphorate*. *Planta Med.* **2004**, *70*, 310–314. [PubMed]
4. Geethangili, M.; Fang, S.H.; Lai, C.H.; Rao, Y.K.; Lien, H.M.; Tzeng, Y.M. Inhibitory effect of *Antrodia camphorata* constituents on the *Helicobacter pylori*-associated gastric inflammation. *Food Chem.* **2010**, *119*, 149–153. [CrossRef]
5. Lai, C.I.; Chu, Y.L.; Ho, C.T.; Su, Y.C.; Kuo, Y.H.; Sheen, L.Y. Antcin K, an active triterpenoid from the fruiting bodies of basswood cultivated *Antrodia cinnamomea*, induces mitochondria and endoplasmic reticulum stress-mediated apoptosis in human hepatoma cells. *J. Tradit. Complement. Med.* **2016**, *6*, 48–56. [CrossRef] [PubMed]

6.  Huang, Y.L.; Chu, Y.L.; Ho, C.T.; Chung, J.G.; Lai, C.I.; Su, Y.C.; Kuo, Y.H.; Sheen, L.Y. Antcin K, an active triterpenoid from the fruiting bodies of basswood cultivated *Antrodia cinnamomea*, inhibits metastasis via suppression of integrin-mediated adhesion, migration, and invasion in human hepatoma cells. *J. Agric. Food Chem.* **2015**, *63*, 4561–4569. [CrossRef] [PubMed]

7.  Tien, A.J.; Chien, C.Y.; Chen, Y.H.; Lin, L.C.; Chien, C.T. Fruiting bodies of *Antrodia cinnamomea* and its active triterpenoid, antcin K, ameliorates *N*-nitrosodiethylamine-induced hepatic inflammation, fibrosis and carcinogenesis in rats. *Am. J. Chin. Med.* **2017**, *45*, 173. [CrossRef] [PubMed]

8.  Kuo, Y.H.; Lin, C.H.; Shih, C.C.; Yang, C.S. Antcin K, a triterpenoid compound from *Antrodia camphorate*, displays antidiabetic and antihyperlipidemic effects via glucose transporter 4 and AMP-activated protein kinase phosphorylation in muscles. *Evid. Based Complement. Alternat. Med.* **2016**. [CrossRef] [PubMed]

9.  Muffler, K.; Leipold, D.; Scheller, M.C.; Haas, C.; Steingroewer, J.; Bley, T.; Neuhaus, H.E.; Mirata, M.A.; Schrader, J.; Ulber, R. Biotransformation of triterpenes. *Process Biochem.* **2011**, *46*, 1–15. [CrossRef]

10. Wang, Y.; Sun, Y.; Wang, C.; Huo, X.; Liu, P.; Wang, C.; Zhang, B.; Zhan, L.; Zhang, H.; Deng, S.; et al. Biotransformation of 11-keto-β-boswellic acid by *Cunninghamella blakesleana*. *Phytochemistry* **2013**, *96*, 330–336. [CrossRef] [PubMed]

11. Sun, Y.; Liu, D.; Xi, R.G.; Wang, X.; Wang, Y.; Hou, J.; Zhang, B.; Wang, X.; Liu, K.; Ma, X. Microbial transformation of acetyl-11-keto-β-boswellic acid and their inhibitory activity on LPS-induced NO production. *Bioorg. Med. Chem. Lett.* **2013**, *23*, 1338–1342. [CrossRef] [PubMed]

12. Wang, W.W.; Xu, S.H.; Zhao, Y.Z.; Zhang, C.; Zhang, Y.Y.; Yu, B.Y.; Zhang, J. Microbial hydroxylation and glycosylation of pentacyclic triterpenes as inhibitors on tissue factor procoagulant activity. *Bioorg. Med. Chem. Lett.* **2017**, *27*, 1026–1030. [CrossRef] [PubMed]

13. Martinez, A.; Rivas, F.; Perojil, A.; Parra, A.; Garcia-Granados, A.; Fernandez-Vivas, A. Biotransformation of oleanolic and maslinic acids by *Rhizomucor miehei*. *Phytochemistry* **2013**, *94*, 229–237. [CrossRef] [PubMed]

14. Qian, L.W.; Zhang, J.; Liu, J.H.; Yu, B.Y. Direct microbial-catalyzed asymmetric α-hydroxylation of betulinic acid by *Nocardia* sp. NRRL 5646. *Tetrahedron Lett.* **2009**, *50*, 2193–2195. [CrossRef]

15. Feng, X.; Luan, J.; Guo, F.F.; Li, D.P.; Chu, Z.Y. Microbial transformation of maslinic acid by *Cunninghamella blakesleana*. *J. Mol. Catal. B Enzym.* **2012**, *82*, 127–130. [CrossRef]

16. Huang, F.X.; Yang, W.Z.; Ye, F.; Tian, J.Y.; Hu, H.B.; Feng, L.M.; Guo, D.A.; Ye, M. Microbial transformation of ursolic acid by *Syncephalastrum racemosum* (Cohn) Schroter AS 3.264. *Phytochemistry* **2012**, *82*, 56–60. [CrossRef] [PubMed]

17. Fu, S.B.; Yang, J.S.; Cui, J.L.; Sun, D.A. Biotransformation of ursolic acid by *Syncephalastrum racemosum* CGMCC 3.2500 and anti-HCV activity. *Fitoterapia* **2013**, *86*, 123–128. [CrossRef] [PubMed]

18. Zhang, C.; Xu, S.H.; Ma, B.L.; Wang, W.W.; Yu, B.Y.; Zhang, J. New derivatives of ursolic acid through the biotransformation by *Bacillus megaterium* CGMCC 1.1741 as inhibitors on nitric oxide production. *Bioorg. Med. Chem. Lett.* **2017**, *27*, 2575–2578. [CrossRef] [PubMed]

19. Liu, X.; Qiao, L.; Xie, D.; Zhang, Y.; Zou, J.; Chen, X.; Dai, J. Microbial transformation of ginsenoside-Rg1 by *Absidia coerulea* and the reversal activity of the metabolites towards multi-drug resistant tumor cells. *Fitoterapia* **2011**, *82*, 1313–1317. [CrossRef] [PubMed]

20. Chen, G.; Yang, M.; Nong, S.; Yang, X.; Ling, Y.; Wang, D.; Wang, X.; Zhang, W. Microbial transformation of 20(S)-protopanaxadiol by *Absidia corymbifera*. Cytotoxic activity of the metabolites against human prostate cancer cells. *Fitoterapia* **2013**, *84*, 6–10. [CrossRef] [PubMed]

21. Chen, G.; Yang, X.; Li, J.; Ge, H.; Song, Y.; Ren, J. Biotransformation of 20(*S*)-protopanaxadiol by *Aspergillus niger* AS 3.1858. *Fitoterapia* **2013**, *91*, 256–260. [CrossRef] [PubMed]

22. Schmitz, D.; Zapp, J.; Bernhardt, R. Hydroxylation of the triterpenoid dipterocarpol with CYP106A2 from *Bacillus megaterium*. *FEBS J.* **2012**, *279*, 1663–1674. [CrossRef] [PubMed]

23. Nei, M.; Kumar, S. *Molecular Evolution and Phylogenetics*; Oxford University Press: New York, NY, USA, 2000; ISBN 100-1-95-13585-7.

24. Kumar, S.; Stecher, G.; Tamura, K. MEGA7: Molecular evolutionary genetics analysis version 7.0 for bigger datasets. *Mol. Biol. Evol.* **2016**, *33*, 1870–1874. [CrossRef] [PubMed]

25. Huang, Y.; Lin, X.; Qiao, X.; Ji, S.; Liu, K.; Yeh, C.; Tzeng, Y.M.; Guo, D.; Ye, M. Antcamphins A–L, ergostanoids from *Antrodia camphorate*. *J. Nat. Prod.* **2014**, *77*, 118–124. [CrossRef] [PubMed]

26. Kouzi, S.A.; Chatterjee, P.; Pezzuto, J.M.; Hamann, M.T. Microbial transformations of the antimelanoma agent betulinic acid. *J. Nat. Prod.* **2000**, *63*, 1653–1657. [CrossRef] [PubMed]

27.  Wang, K.C.; Wang, P.H.; Lee, S.S. Microbial transformation of protopanaxadiol and protopanaxatriol derivatives with *Mycobacterium* sp. (NRRL B-3805). *J. Nat. Prod.* **1997**, *60*, 1236–1241. [CrossRef]

28.  Feng, X.; Zou, Z.; Fu, S.; Sun, L.; Su, Z.; Sun, D.A. Microbial oxidation and glucosidation of echinocystic acid by *Nocardia corallina*. *J. Mol. Catal. B Enzym.* **2010**, *66*, 219–223. [CrossRef]

29.  Grishko, V.V.; Tarasova, E.V.; Ivshina, I.B. Biotransformation of botulin to betulone by growing and resting cells of the actinobacterium *Rhodococcus rhodochrous* IEGM 66. *Process Biochem.* **2013**, *48*, 1640–1644. [CrossRef]

30.  Qiao, X.; Song, W.; Wang, Q.; Liu, K.D.; Zhang, Z.X.; Bo, T.; Li, R.Y.; Liang, L.N.; Tzeng, Y.M.; Guo, D.A.; et al. Comprehensive chemical analysis of triterpenoids and polysaccharides in the medicinal mushroom *Antrodia cinnamomea*. *RSC Adv.* **2015**, *5*, 47040–47052. [CrossRef]

31.  Li, Z.W.; Kuang, Y.; Tang, S.N.; Li, K.; Huang, Y.; Qiao, X.; Yu, S.W.; Tzeng, Y.M.; Lo, J.Y.; Ye, M. Hepatoprotective activities of *Antrodia camphorata* and its triterpenoid compounds against CCl$_4$-induced liver injury in mice. *J. Ethnopharm.* **2017**, *206*, 31–39. [CrossRef] [PubMed]

32.  Bertani, G. Studies on lysogenesis. I. The mode of phage liberation by lysogenic *Escherichia coli*. *J. Bacteriol.* **1951**, *62*, 293–300. [PubMed]

33.  Hugenholtz, P.; Goebel, B.M. *Environmental Molecular Microbiology: Protocols and Applications*; Horizon Scientific Press: Norfolk, UK, 2001; pp. 31–42. ISBN 101-8-98-486-29-8.

34.  Thompson, J.D.; Gibson, T.J.; Higgins, D.G. Multiple sequence alignment using ClustalW and ClustalX. *Curr. Protoc. Bioinform.* **2002**. [CrossRef]

35.  Felsenstein, J. Confidence limits on phylogenies: An approach using the bootstrap. *Evolution* **1985**, *39*, 783–791. [CrossRef] [PubMed]

36.  Posada, D. Model test server: A web-based tool for the statistical selection of models of nucleotide substitution online. *Nucleic Acids Res.* **2006**, *34*, W700–W703. [CrossRef] [PubMed]

37.  Chiang, C.M.; Chang, Y.J.; Wu, J.Y.; Chang, T.S. Production and anti-melanoma activity of methoxyisoflavones from biotransformation of genistein by two recombinant *Escherichia coli*. *Molecules* **2017**, *22*, 87. [CrossRef] [PubMed]

**MDPI**

*Review*

# Total Synthesis and Biological Evaluation of Phaeosphaerides

**Kenichi Kobayashi \*, Kosaku Tanaka III and Hiroshi Kogen**

Graduate School of Pharmaceutical Sciences, Meiji Pharmaceutical University, 2-522-1 Noshio, Kiyose, Tokyo 204-8588, Japan; tanaka.k.cx@m.titech.ac.jp (K.T.); hkogen@my-pharm.ac.jp (H.K.)
\* Correspondence: kenichik@my-pharm.ac.jp; Tel.: +81-42-495-8633

Received: 26 April 2018; Accepted: 7 May 2018; Published: 14 May 2018

**Abstract:** This article reviews studies regarding the total synthesis of phaeosphaerides A and B, nitrogen-containing bicyclic natural products isolated from an endophytic fungus. Numerous synthetic efforts and an X-ray crystal structure analysis of phaeosphaeride A have enabled revision of its originally proposed structure. In addition, a successful protic acid-mediated transformation of phaeosphaeride A to phaeosphaeride B revealed the hypothetical biosynthesis of phaeosphaeride B from phaeosphaeride A. Structure–activity relationship studies of phaeosphaeride derivatives are also discussed.

**Keywords:** phaeosphaeride A; phaeosphaeride B; total synthesis; structural revision; STAT3; anticancer

## 1. Introduction

Signal transducer and activator of transcription 3 (STAT3), which belongs to the STAT family of proteins [1], regulates cell proliferation, differentiation, and survival [2]. Non-activated STAT3 is generally localized in the cytoplasm. Once phosphorylated at Tyr705 in the Janus kinase (JAK)/STAT signaling pathway, STAT3 dimerizes, translocates to the nucleus, and binds to a target DNA sequence to induce transcriptional activation [3,4].

Unusual activation of STAT3 is frequently found in various types of tumor cells, leading to apoptosis resistance and tumor cell proliferation via enhanced expression of gene encoding proteins such as Bcl-2, Bcl-xL, and cyclin D1 [5,6]. Therefore, STAT3 has gained considerable interest as a potential target for anticancer therapy. In fact, various STAT3 inhibitors including synthetic small molecules and natural products have been evaluated as prospective anticancer chemotherapeutic agents [7].

In 2006, phaeosphaerides A (proposed structure **1a**) was isolated from the endophytic fungus FA39 (*Phaeosphaeria avenaria*) by Clardy and co-workers as an inhibitor of STAT3-DNA binding (Figure 1) [8]. Phaeosphaeride A possesses a bicyclic structure with three contiguous stereocenters in its dihydropyran ring. Phaeosphaeride A inhibits STAT3 activity with an $IC_{50}$ of 0.61 mM and also inhibits cell growth in STAT3-dependent U266 multiple myeloma cells with an $IC_{50}$ of 6.7 μM. Therefore, phaeosphaeride A is expected to be a potential lead compound for anticancer drug candidates. On the other hand, phaeosphaeride B (**1b**), the C-8 stereoisomer of phaeosphaeride A, was reported to have no STAT3 inhibitory activity.

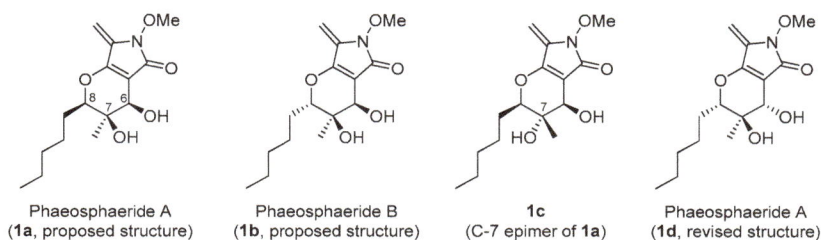

**Figure 1.** Structures of phaeosphaerides.

The promising biological activity of phaeosphaeride A as well as its simple and unique molecular structure has attracted much attention from the synthetic community. In addition, structure–activity relationship (SAR) studies of STAT3 inhibitory activity should be of significant importance for potential anticancer therapy.

## 2. Synthetic Approach toward Phaeosphaeride A

### 2.1. Synthesis of the Proposed Structure of Phaeosphaeride A

In 2011, our group accomplished the first total synthesis of the proposed structure of phaeosphaeride A (**1a**; Schemes 1 and 2) [9]. In this synthesis, to construct the C-7 and C-8 stereocenters, we used a typical *E*-selective Horner–Wadsworth–Emmons reaction followed by Sharpless asymmetric dihydroxylation using AD-mix-β to obtain the requisite diol (2*S*,3*R*)-**4** in high yield. In this step, the absolute configuration and the high enantiomeric excess of diol **4** were confirmed by a modified Mosher's method. After appropriate conversion of diol **4** to secondary alcohol **5**, we found good reaction conditions for an oxy-Michael addition of alcohol **5** to dimethyl acetylenedicarboxylate: Use of a catalytic amount of *n*-BuLi cleanly afforded the desired Michael adduct (*E*)-**6** along with the (*Z*)-isomer in 76% and 18% yields, respectively. The remaining C-6 stereogenic center was installed by a vinyl-anion aldol reaction of aldehyde **7** using sodium bis(trimethylsilyl)amide (NaHMDS), providing the desired dihydropyran derivative **8** via the plausible transition state shown in Scheme 1.

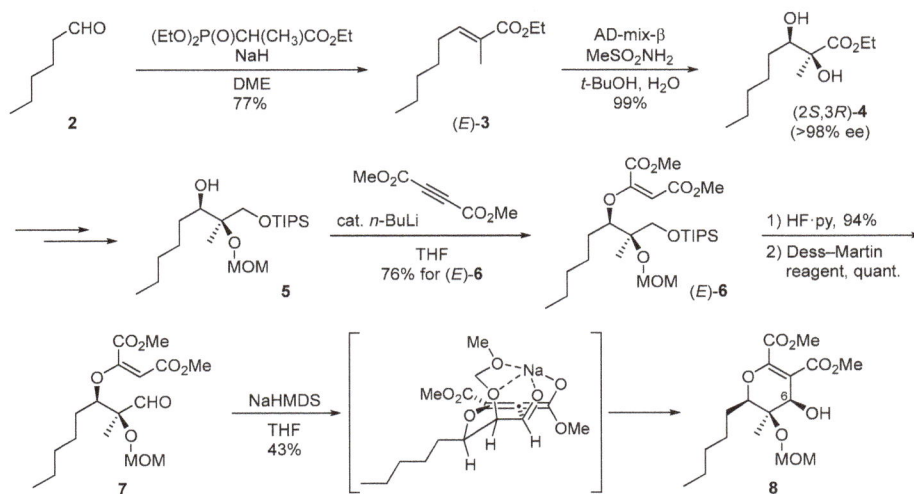

**Scheme 1.** Synthesis of the dihydropyran intermediate **8**.

The construction of the five-membered ring in phaeosphaeride A was followed by regioselective installation of the exo-methylene group to furnish the proposed structure of phaeosphaeride A (**1a**; Scheme 2). However, the $^1$H and $^{13}$C NMR spectra of synthetic **1a** did not match those reported for natural phaeosphaeride A. Therefore, this synthesis revealed that the structure of phaeosphaeride A had been incorrectly assigned.

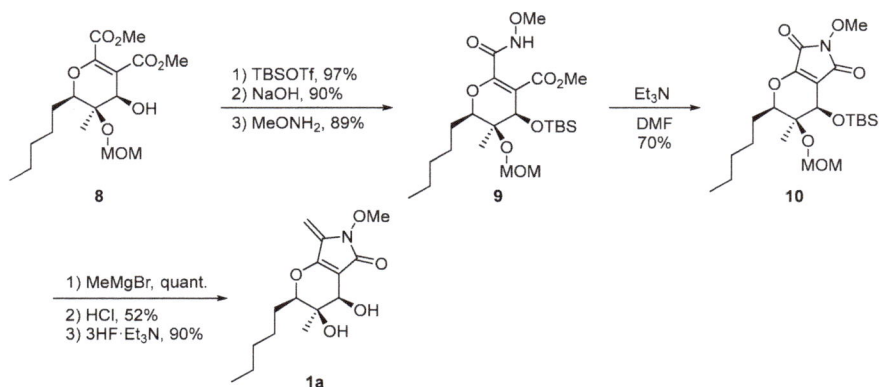

**Scheme 2.** Tamura's total synthesis of the proposed structure of phaeosphaeride A (**1a**).

For natural phaeosphaeride A, the Clardy group observed nuclear Overhauser effect spectroscopy (NOESY) correlations between H-6 and H-8, and between H-15 and both H-6 and H-8 (Figure 2) [8]. The correlation between H-6 and H-8 clearly indicates a pseudodiaxial relationship between these hydrogens. However, the correlations between H-15 and both H-6 and H-8 do not provide information about the configuration of the C-7 stereocenter. Hence, the correct structure of phaeosphaeride A was presumed to be the C-7 epimer **1c** of the originally proposed structure, or the epimer's enantiomer **1d**.

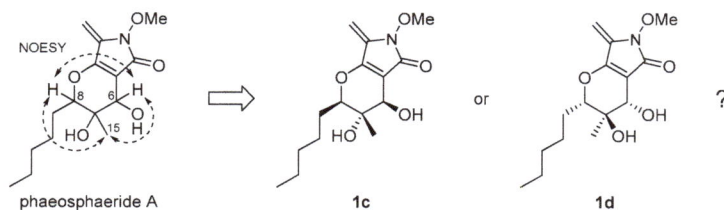

**Figure 2.** Results of NOESY experiments by Clardy.

In 2012, Sarli's group also succeeded in the total synthesis of the proposed structure of phaeosphaeride A (**1a**) [10]. Their synthesis involved a strategy similar to ours, in which dihydroxylation of unsaturated ester (*E*)-**11** using catalytic OsO$_4$ and 4-methylmorpholine *N*-oxide (NMO) was used to form the C-7 and C-8 stereocenters. Then, the C-6 center was stereochemically controlled by the anti-selective addition of a vinyllithium species to aldehyde **13** via the Felkin–Ahn transition state. After establishing the three contiguous stereocenters, sequential oxy-Michael addition/methanol elimination followed by selective dehydration furnished (±)-**1a** (Scheme 3).

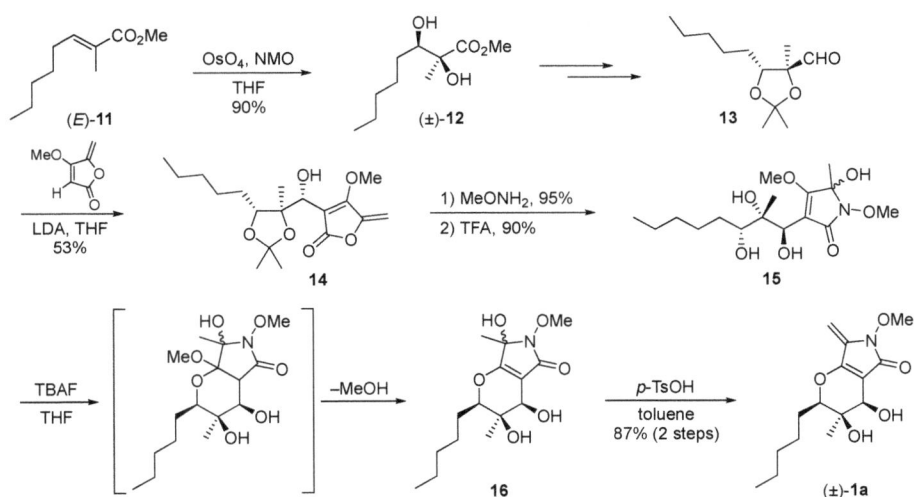

**Scheme 3.** Sarli's total synthesis of (±)-**1a**.

They also synthesized both enantiomers of **1a**, (6*R*,7*R*,8*R*)-**1a** and (6*S*,7*S*,8*S*)-**1a**, by Sharpless asymmetric dihydroxylation of (*E*)-**11** in the first step using AD-mix-β and AD-mix-α, respectively. They obtained the crystal structure of synthetic (6*R*,7*R*,8*R*)-**1a** using synchrotron radiation, which proved their structural assignment of **1a** by NMR. Their studies also pointed to the structural revision of phaeosphaeride A to **1c** or its enantiomer **1d** (Scheme 4).

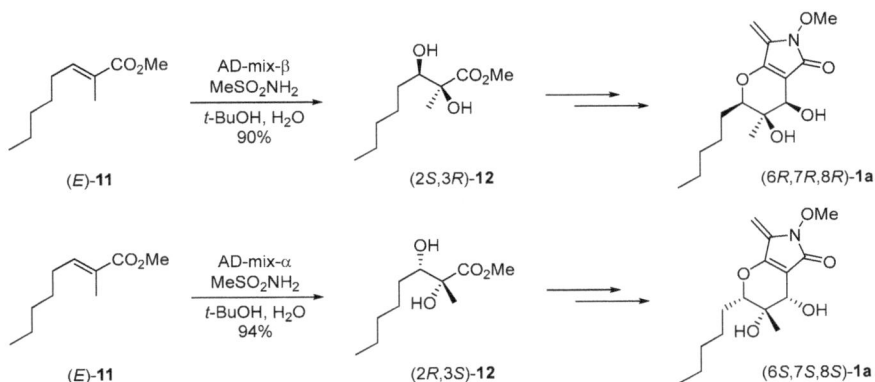

**Scheme 4.** Sarli's total synthesis of (6*R*,7*R*,8*R*)-**1a** and (6*S*,7*S*,8*S*)-**1a**.

## 2.2. Stereochemical Determination of Natural (−)-Phaeosphaeride A

Based on the above achievements by our group and Sarli's group, the C-7 epimer **1c** of the originally proposed structure or the epimer's enantiomer **1d** needed to be synthesized to resolve the issue of the stereochemistry of natural phaeosphaeride A. To access **1c** by a synthetic strategy similar to those shown in Schemes 1 and 2, (*Z*)-α,β-unsaturated ester (*Z*)-**3** was first prepared instead of the (*E*)-ester by Still–Gennari olefination using phosphonate **17**. The ester (*Z*)-**3** was converted into the diol (2*R*,3*R*)-**18** via Sharpless dihydroxylation using AD-mix-β. According to the previous route with slight modification, the intermediate diol (2*R*,3*R*)-**18** was successfully converted via dihydropyran intermediate **19** into **1c**, the $^{1}$H and $^{13}$C NMR spectra of which completely matched the literature data for natural phaeosphaeride A, and the optical rotation of the synthetic compound had the opposite sign to that of the natural product. The correct structure of natural (−)-phaeosphaeride A was thus shown to be the enantiomer **1d** of synthetic **1c** [11]. Then, natural (−)-phaeosphaeride A was synthesized by using AD-mix-α instead of AD-mix-β in the first step (Scheme 5) [12].

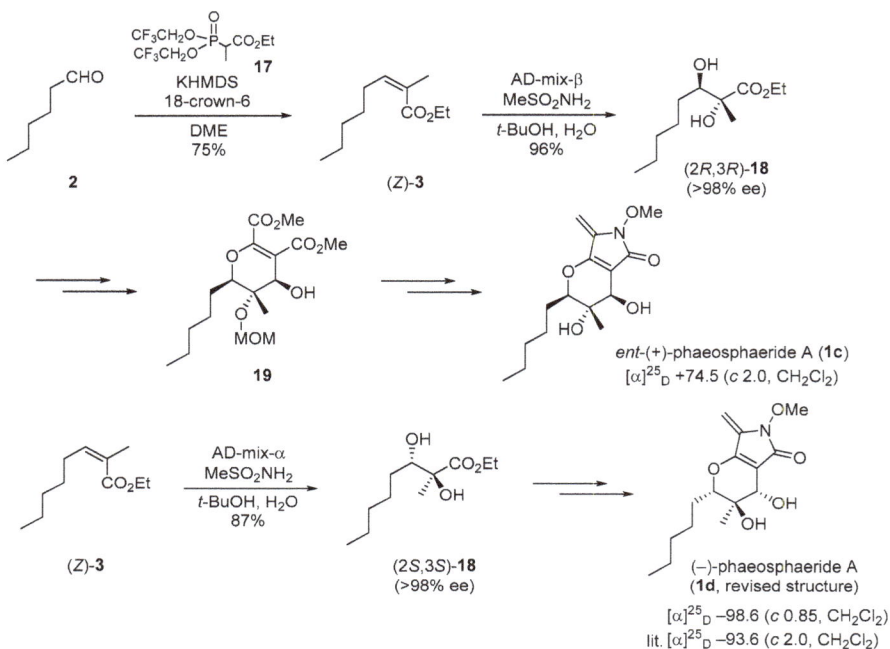

**Scheme 5.** Kobayashi and Kogen's total synthesis of **1c** and **1d**.

After the reports, Abzianidze et al. reported the crystal structure of natural phaeosphaeride A [13]. This crystal structure clearly supported the results of stereochemical revision by these synthetic approaches.

## 3. Synthetic Approach toward Phaeosphaeride B

The first total synthesis of (±)-phaeosphaeride B (**1b**) was reported by Sarli's group in 2014 [14]. Their synthetic strategy for **1a** (Scheme 3) could be applied to the preparation of (±)-**1b**. In their synthesis, conversion of (*E*)-**11** to an allylic alcohol followed by epoxidation and subsequent regioselective Ti(O-*i*Pr)$_4$-mediated epoxide ring-opening of **20** with allyl alcohol enabled introduction of the C-7 and C-8 stereocenters of (±)-**1b**. The key anti-selective nucleophilic addition of a vinyllithium species to aldehyde **22** delivered the required C-6 stereocenter through the polar Felkin–Ahn transition state. Then, the lactam and dihydropyran rings and exo-methylene group were assembled to complete the total synthesis of (±)-phaeosphaeride B (**1b**) (Scheme 6).

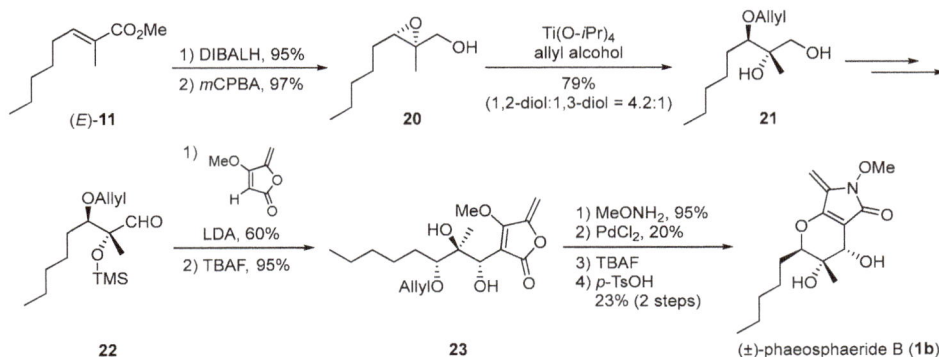

**Scheme 6.** Sarli's total synthesis of (±)-phaeosphaeride B (**1b**).

They also developed an improved synthetic scheme for (±)-**1b**, in which bis-TMS ether aldehyde **25**, derived from diol (±)-**12** via cyclic sulfate **24**, was used in the reaction with α-lithio tetronate to form **26a** along with the TMS-migrated product **26b**. Subsequent conversion via their established route effectively yielded (±)-**1b** (Scheme 7).

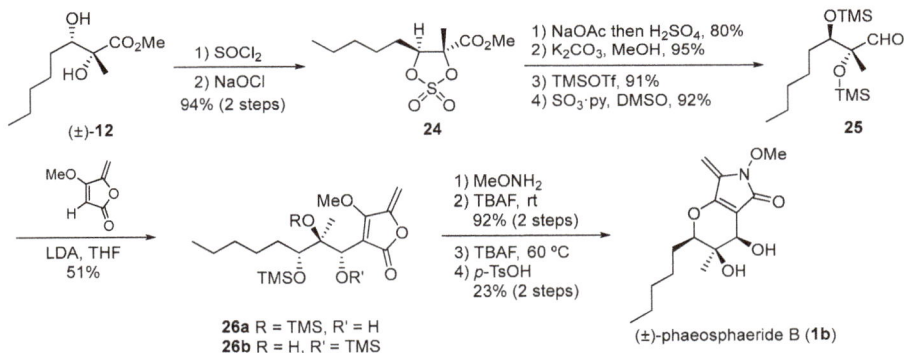

**Scheme 7.** Improved synthesis of (±)-**1b** by Sarli et al.

After the successful total synthesis of (±)-phaeosphaeride B by Sarli, we demonstrated a biomimetic transformation from (−)-phaeosphaeride A to (−)-phaeosphaeride B [12]. We posited that phaeosphaerides A and B would be biosynthetically interconverted under acidic conditions. In testing this hypothesis, treatment of synthetic (−)-phaeosphaeride A (**1d**) with trifluoroacetic acid (TFA) as a protic acid gave the corresponding trifluoroacetate **27** with stereochemical inversion at C-6 stereocenter via dehydrative formation of the oxonium cation intermediate **A**. The labile trifluoroacetate **27** was immediately hydrolyzed with aqueous NaHCO$_3$ in THF to yield (−)-phaeosphaeride B (**1b**) in a good yield (Scheme 8). In addition, this synthesis confirmed the absolute configuration of natural (−)-phaeosphaeride B as shown in Scheme 8.

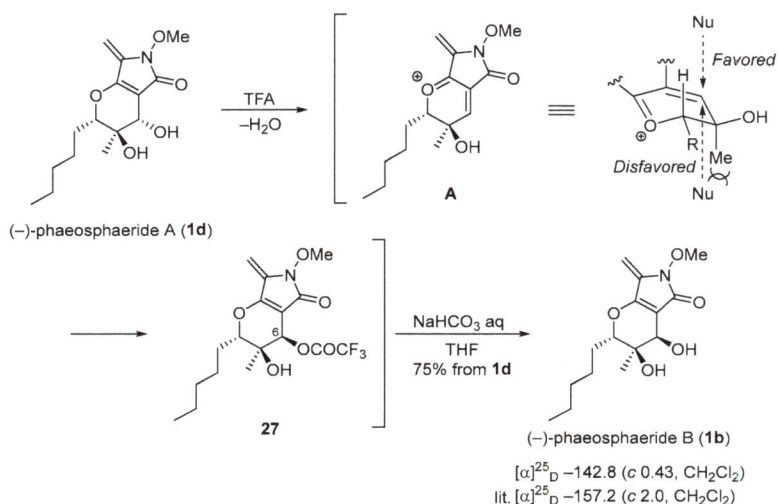

**Scheme 8.** Biomimetic transformation from (−)-phaeosphaeride A to (−)-phaeosphaeride B.

## 4. Biological Evaluation of Phaeosphaerides and Their Derivatives

Considering their potential as a seed compound for anticancer treatment, Sarli's and Abzianidze's groups evaluated the biological activities of phaeosphaerides and their synthetic derivatives [10,14–17].

Initially, Sarli and colleagues biologically evaluated the stereoisomers of phaeosphaeride A, (6*R*,7*R*,8*R*)-**1a** and (6*S*,7*S*,8*S*)-**1a** [10]. These compounds inhibited STAT3-dependent transcriptional activity in a dose-dependent manner and decreased cell proliferation in breast (MDA-MB-231) and pancreatic (PANC-1) cancer cells in the low micromolar range. After that, they reported that the synthetic (6*S*,7*S*,8*S*)-**1a** and (6*R*,7*S*,8*S*)-phaeosphaeride had only very weak inhibitory activity against binding of STAT3 to its phosphotyrosine peptide ligand, suggesting that phaeosphaerides are upstream inhibitors of a tyrosine kinase in the JAK/STAT pathway [14].

Abzianidze et al. prepared and biologically evaluated the C-6 acyl derivatives **28** and **29**, bis-methanol adducts **30a** and **30b** without the MeO group on the nitrogen, and hydrolyzed products **31a** and **31b** prepared from isolated natural phaeosphaeride A [15]. Compared to natural phaeosphaeride A (EC$_{50}$ = 46 ± 5 µM), chloroacetyl derivative **29** exhibited more potent cytotoxicity (EC$_{50}$ = 33 ± 7 µM) against the A549 cancer cell line, while synthetic **30** and **31** had no activity (Scheme 9).

Additionally, they also synthesized 7-(4-methylphenyl)thiomethyl and 7-morpholylmethyl derivatives **32** and **33**, which were less cytotoxic than the parent phaeosphaeride A (Scheme 10) [16]. These results clearly indicated that the exo-methylene and *N*-OMe groups were essential for potent cytotoxicity. Their studies strongly suggest that further SAR studies can provide lead compounds with greater potency for potential use as anticancer chemotherapeutic agents.

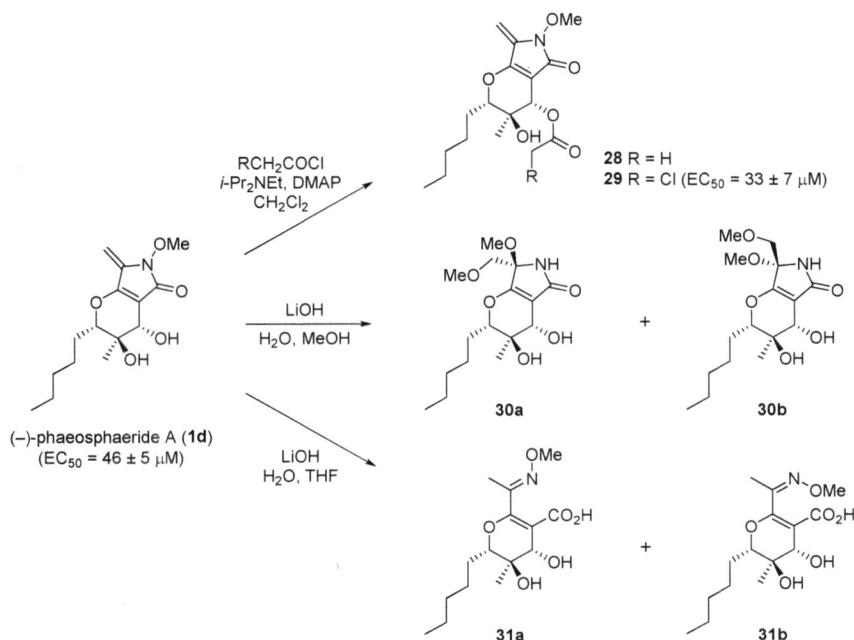

**Scheme 9.** Phaeosphaeride A derivatives **28**, **29**, **30**, and **31** prepared by Abzianidze et al.

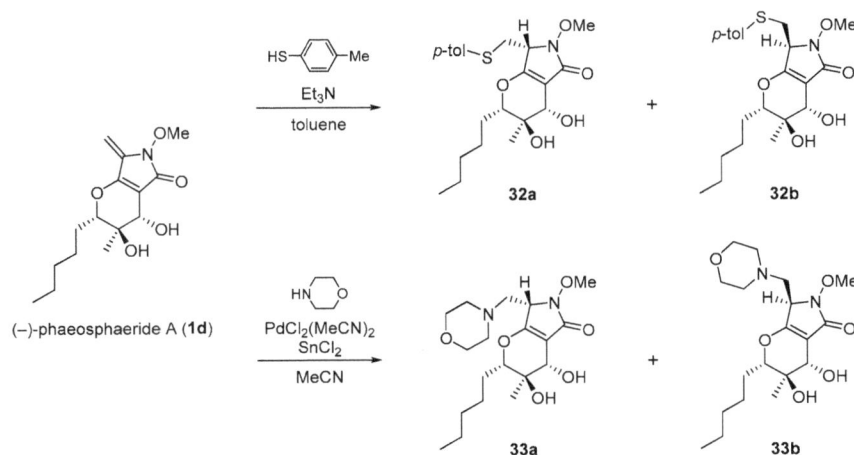

**Scheme 10.** Phaeosphaeride A derivatives **32** and **33** by Abzianidze et al.

## 5. Conclusions

Phaeosphaerides A and B have attracted considerable attention due not only to their chemical structures but also to their biological activity, and their stereochemical structures have been unambiguously determined through the total synthesis and X-ray crystal structure analysis of phaeosphaeride A. The promising anticancer activity of phaeosphaeride A based on inhibition of STAT3-DNA binding indicates that this natural product is a promising seed compound for anticancer drug candidates. SAR studies on phaeosphaerides led to the development of more potent compounds such as chloroacetyl derivative **29**, and further SAR studies are awaited for anticancer drug discovery research. In addition, structurally related natural products including paraphaeosphaerides [18,19], phyllostictines [20,21], isoaigialones [22], benesudon [23], and curvupallides [24] should also attract considerable interest as pharmaceutical targets. Further studies of phaeosphaerides are expected to make valuable contributions to synthetic and medicinal chemistry.

**Author Contributions:** K.K. and K.T.III prepared the manuscript. K.K. managed the project with assistance from H.K.

**Acknowledgments:** This work was supported by JSPS KAKENHI Grant Number 25860015 and partially by a grant from the Dementia Drug Resource Development Center Project S1511016, the Ministry of Education, Culture, Sports Science and Technology (MEXT), Japan.

**Conflicts of Interest:** The authors declare no conflict of interest.

## References

1. Levy, D.E. Physiological significance of STAT proteins: Investigations through gene disruption in vivo. *Cell. Mol. Life Sci.* **1999**, *55*, 1559–1567. [CrossRef] [PubMed]
2. Darnell, J.E., Jr. STATs and gene regulation. *Science* **1997**, *277*, 1630–1635. [CrossRef] [PubMed]
3. Darnell, J.E., Jr.; Kerr, I.M.; Stark, G.R. Jak-Stat pathways and transcriptional activation in response to IFNs and other extracellular signaling proteins. *Science* **1994**, *264*, 1415–1421. [CrossRef] [PubMed]
4. Zhong, Z.; Darnell, J.E., Jr. Stat3: A STAT family member activated by tyrosine phosphorylation in response to epidermal growth factor and interleukin-6. *Science* **1994**, *264*, 95–98. [CrossRef] [PubMed]
5. Bromberg, J.F.; Wrzeszczynska, M.H.; Devgan, G.; Zhao, Y.; Pestell, R.G.; Albanese, C.; Darnell, J.E., Jr. Stat3 as an oncogene. *Cell* **1999**, *98*, 295–303. [CrossRef]
6. Subramaniam, A.; Shanmugam, M.K.; Perumal, E.; Li, F.; Nachiyappan, A.; Dai, X.; Swamy, S.N.; Ahn, K.S.; Kumar, A.P.; Tan, B.K.; et al. Potential role of signal transducer and activator of transcription (STAT)3 signaling pathway in inflammation, survival, proliferation and invasion of hepatocellular carcinoma. *Biochim. Biophys. Acta Rev. Cancer* **2013**, *1835*, 46–60. [CrossRef] [PubMed]
7. Siveen, K.S.; Sikka, S.; Surana, R.; Dai, X.; Zhang, J.; Kumar, A.P.; Tan, B.K.; Sethi, G.; Bishayee, A. Targeting the STAT3 signaling pathway in cancer: Role of synthetic and natural inhibitors. *Biochim. Biophys. Acta Rev. Cancer* **2014**, *1845*, 136–154. [CrossRef] [PubMed]
8. Maloney, K.N.; Hao, W.; Xu, J.; Gibbons, J.; Hucul, J.; Roll, D.; Brady, S.F.; Schroeder, F.C.; Clardy, J. Phaeosphaeride A, an Inhibitor of STAT3-dependent signaling isolated from an endophytic fungus. *Org. Lett.* **2006**, *8*, 4067–4070. [CrossRef] [PubMed]
9. Kobayashi, K.; Okamoto, I.; Morita, N.; Kiyotani, T.; Tamura, O. Synthesis of the proposed structure of phaeosphaeride A. *Org. Biomol. Chem.* **2011**, *9*, 5825–5832. [CrossRef] [PubMed]
10. Chatzimpaloglou, A.; Yavropoulou, M.P.; Rooij, K.E.; Biedermann, R.; Mueller, U.; Kaskel, S.; Sarli, V. Total synthesis and biological activity of the proposed structure of phaeosphaeride A. *J. Org. Chem.* **2012**, *77*, 9659–9667. [CrossRef] [PubMed]
11. Kobayashi, K.; Kobayashi, Y.; Nakamura, M.; Tamura, O.; Kogen, H. Establishment of relative and absolute configurations of phaeosphaeride A: Total synthesis of *ent*-phaeosphaeride A. *J. Org. Chem.* **2015**, *80*, 1243–1248. [CrossRef] [PubMed]
12. Kobayashi, K.; Kunimura, R.; Tanaka, K., III; Tamura, O.; Kogen, H. Total synthesis of (−)-phaeosphaeride B by a biomimetic conversion from (−)-phaeosphaeride A. *Tetrahedron* **2017**, *73*, 2382–2388. [CrossRef]
13. Abzianidze, V.V.; Poluektova, E.V.; Bolshakova, K.P.; Panikorovskii, T.L.; Bogachenkov, A.S.; Berestetskiy, A.O. Crystal structure of natural phaeosphaeride A. *Acta Crystallogr.* **2015**, *E71*, o625–o626. [CrossRef] [PubMed]
14. Chatzimpaloglou, A.; Kolosov, M.; Eckols, T.K.; Tweardy, D.J.; Sarli, V. Synthetic and biological studies of phaeosphaerides. *J. Org. Chem.* **2014**, *79*, 4043–4054. [CrossRef] [PubMed]
15. Abzianidze, V.V.; Prokofieva, D.S.; Chisty, L.A.; Bolshakova, K.P.; Berestetskiy, A.O.; Panikorovskii, T.L.; Bogachenkov, A.S.; Holder, A.A. Synthesis of natural phaeosphaeride A derivatives and an in vitro evaluation of their anti-cancer potential. *Bioorg. Med. Chem. Lett.* **2015**, *25*, 5566–5569. [CrossRef] [PubMed]

16. Abzianidze, V.V.; Bolshakova, K.P.; Prokofieva, D.S.; Berestetskiy, A.O.; Kuznetsova, V.A.; Trishin, Y.G. Synthesis of 7-(4-methylphenyl)thiomethyl and 7-morpholylmethyl derivatives of natural phaeosphaeride A and their cytotoxic activity. *Mendeleev Commun.* **2017**, *27*, 82–84. [CrossRef]

17. Abzianidze, V.V.; Efimova, K.P.; Poluektova, E.V.; Trishin, Y.G.; Kuznetsov, V.A. Synthesis of natural phaeosphaeride A and semi-natural phaeosphaeride B derivatives. *Mendeleev Commun.* **2017**, *27*, 490–492. [CrossRef]

18. Li, C.S.; Ding, Y.; Yang, B.J.; Miklossy, G.; Yin, H.Q.; Walker, L.A.; Turkson, J.; Cao, S. A new metabolite with a unique 4-pyranone−γ-lactam−1,4-thiazine moiety from a Hawaiian-plant associated fungus. *Org. Lett.* **2015**, *17*, 3556–3559. [CrossRef] [PubMed]

19. Li, C.S.; Sarotti, A.M.; Huang, P.; Dang, U.T.; Hurdle, J.G.; Kondratyuk, T.P.; Pezzuto, J.M.; Turkson, J.; Cao, S. NF-κB inhibitors, unique γ-pyranol-γ-lactams with sulfide and sulfoxide moieties from Hawaiian plant *Lycopodiella cernua* derived fungus *Paraphaeosphaeria neglecta* FT462. *Sci. Rep.* **2017**, *7*, 10424–10433. [CrossRef] [PubMed]

20. Evidente, A.; Cimmino, A.; Andolfi, A.; Vurro, M.; Zonno, M.C.; Cantrell, C.L.; Motta, A. Phyllostictines A–D, oxazatricycloalkenones produced by *Phyllosticta cirsii*, a potential mycoherbicide for *Cirsium arvense* biocontrol. *Tetrahedron* **2008**, *64*, 1612–1619. [CrossRef]

21. Trenti, F.; Cox, R.J. Structural revision and biosynthesis of the fungal phytotoxins phyllostictines A and B. *J. Nat. Prod.* **2017**, *80*, 1235–1240. [CrossRef] [PubMed]

22. Silva, G.H.; Zeraik, M.L.; Oliveira, C.M.; Teles, H.L.; Trevisan, H.C.; Pfenning, L.H.; Nicolli, C.P.; Young, M.C.M.; Mascarenhas, Y.P.; Abreu, L.M.; et al. Lactone derivatives produced by a *Phaeoacremonium* sp., an endophytic fungus from *Senna spectabilis*. *J. Nat. Prod.* **2017**, *80*, 1674–1678. [CrossRef] [PubMed]

23. Thines, E.; Arendholz, W.-R.; Anke, H. Bunesudon, a new antibiotic fungal metabolite from cultures of *Mollisia benesuada* (Tul.) Phill. *J. Antibiot.* **1997**, *50*, 13–17. [CrossRef] [PubMed]

24. Abraham, W.-R.; Meyer, H.; Abate, D. Curvupallides, a new class of alkaloids from the fungus *Curvularia pallescens*. *Tetrahedron* **1995**, *51*, 4947–4952. [CrossRef]

MDPI

St. Alban-Anlage 66

4052 Basel

Switzerland

Tel. +41 61 683 77 34

Fax +41 61 302 89 18

www.mdpi.com

*Catalysts* Editorial Office

E-mail: catalysts@mdpi.com

www.mdpi.com/journal/catalysts

www.ingramcontent.com/pod-product-compliance
Lightning Source LLC
Chambersburg PA
CBHW051917210326

41597CB00033B/6172